Democracy in Practice

Public Participation in Environmental Decisions

Thomas C. Beierle
and Jerry Cayford

Resources for the Future
Washington, DC

Printed in the United States of America

An RFF Press book
Published by Resources for the Future
1616 P Street, NW, Washington, DC 20036–1400
www.rff.org

Library of Congress Cataloging-in-Publication Data
Beierle, Thomas C.
    Democracy in practice : public participation in environmental decisions / Thomas C. Beierle and Jerry Cayford.
        p. cm.
    Includes bibliographical references and index.
    ISBN 1-891853-53-8 (lib. bdg.) — ISBN 1-891853-54-6 (pbk.)
        1. Environmental policy—United States—Citizen participation—Case studies.
    I. Cayford, Jerry. II. Title
GE180 .B45 2002
363.7′0525—dc21   2002017322

f e d c b a

The paper in this book meets the guidelines for permanence and durability of the Committee on Production Guidelines for Book Longevity of the Council on Library Resources.

The text of this book was designed and typeset by Betsy Kulamer in Stone Serif and Stone Sans. It was copyedited by Pamela D. Angulo. The cover was designed by Rosenbohm Graphic Design.

ISBN 1–891853–53–8 (cloth)   ISBN 1–891853–54–6 (paper)

About
Resources for the Future
and *RFF Press*

Resources for the Future (RFF) improves environmental and natural resource policymaking worldwide through independent social science research of the highest caliber.

Founded in 1952, RFF pioneered the application of economics as a tool to develop more effective policy about the use and conservation of natural resources. Its scholars continue to employ social science methods to analyze critical issues concerning pollution control, energy policy, land and water use, hazardous waste, climate change, biodiversity, and the environmental challenges of developing countries.

RFF Press supports the mission of RFF by publishing book-length works that present a broad range of approaches to the study of natural resources and the environment. Its authors and editors include RFF staff, researchers from the larger academic and policy communities, and journalists. Audiences for RFF publications include all of the participants in the policymaking process—scholars, the media, advocacy groups, nongovernmental organizations, professionals in business and government, and the general public.

Resources for the Future

Contents

Acknowledgements

This research owes much to Terry Davies, a senior fellow at Resources for the Future (RFF), who encouraged us to think beyond doing yet another case study on public participation. Terry suggested the idea of a meta-analysis of cases, and he deserves much thanks for his invaluable ideas and advice throughout the project.

Bringing the project from idea to reality required long hours of hunting for case studies and then reading them with much more than the usual attention to detail. Anne McEnany spent the summer of 1999 identifying case studies and tracking them down. Anne's work was aided by RFF librarian Chris Clotworthy, whose own investigative skills were more than adequate to the task. Particular thanks go to David Konisky, who not only helped us code these case studies but also helped test and refine the coding methodology during the pilot project. We could not have pulled together data on 239 cases without his contributions. Finally, we are grateful to all the authors of the case studies for producing the rich body of literature on which our research is based.

Several reviewers provided insightful suggestions as our research proceeded. Julia Abelson, Patricia Bonner, Deborah Dalton, Dan Fiorino, Christopher Foreman, Achim Halpaap, Troy Hartley, Tom Webler, and four anonymous reviewers graciously read and commented on the manuscript that arrived in their mailboxes. Juliana Birkhoff, Gail Charnley, Caron Chess, and participants in a workshop on conflict resolution provided helpful feedback on papers that covered various parts of the research. At RFF, Allen Blackman, Terry Davies, Bob Hersh, Jim Sanchirico, Mike Toman, and visiting scholar Allan Mazur provided assistance and comments on various aspects of the project. The final manuscript benefited greatly from the stewardship of Felicia Day and Don Reisman, detailed editing by Pamela Angulo and Sally

Atwater, and the research assistance of Maria Reff. Many other colleagues inside and outside of RFF contributed their ideas and suggestions after presentations of the research at seminars, workshops, and conferences over the past two years.

We appreciate that many people made this book a better product by giving so freely of their time. Not all will agree with our interpretations, conclusions, and recommendations, which—along with all errors—remain our responsibility.

For the cover art on the paperback edition of this book, we would like to thank Steve Kinney for the background picture of the Sempra Power Plant public hearing in New Milford, Connecticut, and Jeff Leard for the two inset pictures.

This material is based upon work supported by the National Science Foundation under Grant No. 9818728. Any opinions, findings, and conclusions or recommendations expressed in this material are those of the authors and do not necessarily reflect the views of the National Science Foundation.

Chapter 1

Introduction

O ver the past 30 years, public participation has taken center stage in the play of influences that determine how society will manage and protect the environment. Its increasing role in environmental policymaking has led to much recent discussion—accompanied by some cheering, some hand-wringing, a great deal of speculation, and always a recognition of its growing importance.

A broad array of processes that emphasize face-to-face deliberation, problem-solving, and consensus building have joined traditional public hearings and public comment procedures. Policy dialogues, stakeholder advisory committees, citizen juries, facilitated mediations, and various other processes are now familiar components of the public participation mix. The amount of influence the public can wield has changed as well. Agreements in regulatory negotiations among interest groups, for example, actually determine the content of proposed environmental rules. Understanding the role of public participation is increasingly crucial for understanding how government makes and carries out environmental policy.

Over the past three decades, the thousands of cases in which the public has become involved in U.S. environmental policy decisions have produced many hundreds of documents that describe what happened in one case or another. In this book, we provide a systematic analysis of this case study literature and evaluate how well public participation has performed in its central role in environmental policymaking. We synthesize the work of hundreds of researchers to describe what public participation has accomplished and to understand what leads to effective processes.

Two main objectives guided our research. First and foremost, we wanted to develop an understanding of the social value of public participation, that is, its "value added" for society. To do so, we identified several "social goals" for public participation and used these goals as criteria for evaluating the success of public participation efforts. Second, we wanted to understand what makes some processes successful and others not. We examined how the success of public participation varies with the kinds of issues under debate and other aspects of the context in which participation takes place. We also compared the effectiveness of several different approaches to public participation, from public hearings to formal mediations. The literature is full of hypotheses about what makes public participation successful, and data from the case study record allowed us to test some of these hypotheses empirically.

Our study is the first to apply a consistent evaluation framework to such a large, diverse set of public participation cases. It joins a small but growing body of research that draws lessons from numerous cases of particular types of participation, such as environmental mediation (Bingham 1986), collaborative watershed partnerships (Leach, Pelkey, and Sabatier 2000), and regulatory negotiations (Coglianese 1997). However, the bulk of the previous literature on public participation analyzes one case study or a few case studies. Such analyses are not comprehensive enough to facilitate understanding of how the practice of public participation changes from one decisionmaking context to another or from one method of participation to another. Previous studies have used widely varying definitions of success, making it difficult to compare their conclusions or extrapolate beyond a given study's focus. Several researchers have noted the need to draw more broadly applicable lessons from the experience of public participation (Kraft 1995; Lynn and Busenberg 1995; NNOTF 1997; Stern and Fineberg 1996). This book responds to these calls.

A Brief History of Public Participation in Environmental Decisionmaking

Public participation is best understood as a challenge to the traditional management of government policy by experts in administrative agencies. From the late nineteenth century until the middle of the twentieth century, public administration in the United States was dominated by the "managerial" model in which government administrators were entrusted to identify and pursue the common good.

As government responsibilities increased in scope and complexity, large professional bureaucracies grew to manage them. Around the turn of the century, for example, Gifford Pinchot established a strong managerial ethos for the nascent U.S. Forest Service, which was responsible for managing vast tracts of public land. Through "scientific forestry," Pinchot sought to serve the public interest by applying conservation policies that produced the greatest good for the greatest number for the longest time (Hays 1959). Such a concept of social welfare maximization still drives managerialism. It is most often associated in modern environmental policymaking with the decision tools of risk assessment and cost–benefit analysis (Breyer 1995).

The managerial approach to decisionmaking has always occupied a tenuous position in democracy. In the context of regulatory rulemaking, for example, the demands of expert management and the demands of democratic accountability can be at odds with each other:

> On the one hand, we have established that in order for government to be truly responsive to the incessant demands of the American people for public programs to solve private problems, rulemaking is essential.... On the other hand, as an indispensable surrogate to the legislative process, rulemaking has a fundamental flaw that violates basic democratic principles. Those who write the law embodied in rules are not elected; they are accountable to the American people only through indirect and less than foolproof means. (Kerwin 1999, 157)

A fundamental challenge for administrative governance is reconciling the need for expertise in managing administrative programs with the transparency and participation demanded by a democratic system. Kerwin (1999) argues that major expansions of government programs have brought this tension between managerialism and public accountability to a head at various times in the nation's history, each time to be relieved by new legislation promoting public participation.

The first example is the New Deal of the 1930s, which dramatically increased the influence of the executive branch of government in the workings of the economy. It provoked legislative reaction in the form of the Administrative Procedure Act (APA) in 1946. The APA systematized for the first time the process that federal agencies must use when making law through rulemaking. It requires that agencies provide public notice about the rules they are proposing, information on which the rules are based, an opportunity for public comment on those rules, and judicial review of the rulemaking process. The APA continues to govern all regulatory proceedings and is the cornerstone of public participation in administrative governance.

The decades after the passage of the APA saw increasing challenges to the managerial model. Traditional tensions between expertise and accountability were exacerbated by increasing skepticism that managers could adequately identify a public interest in ever-more-complicated administrative systems. The concept of pluralism began to replace managerialism as the dominant paradigm of administrative decisionmaking (Stewart 1975; Reich 1985). According to the pluralist view, government administrators were not a source of objective decisionmaking in the public interest but arbiters among different interests within the public. Whereas the managerial perspective identified maximization of social welfare as the ultimate social goal, pluralism did not recognize an objective sense of the "public good." Rather, a contingent public good was to be debated and arrived at by negotiation among interests (Williams and Matheny 1995).

Pluralism flourished in a burst of public participation legislation that followed the expansion of government under President Lyndon Johnson's Great Society programs in the 1960s. From 1966 to 1976, the U.S. Congress passed the Freedom of Information Act (1966), the Federal Advisory Committee Act (1972), the Privacy Act (1974), and the Government in the Sunshine Act (1976). Together, these laws greatly expanded citizens' access to government information and decisionmaking. The major environmental statutes of the 1970s were also "pluralist-created and pluralist-driven," with their strong provisions for public review and citizen suits, which gave interest groups bargaining rights with industry (Gauna 1998, 24). During the past three decades, interest group membership and representation in Washington have risen dramatically to take advantage of the pluralism enshrined in environmental laws (Coglianese 1999b).

In recent years, the pluralist paradigm has come under pressure from an even more intensely participatory perspective. This "popular" democratic theory stresses the importance of the act of participation, not only in influencing decisions but also in strengthening civic capacity and social capital. Like pluralism, popular democracy emphasizes interaction among often adversarial interests, but that interaction is viewed less as a competitive negotiation than as a way to identify the common good and subsequently act on shared communal (versus individual) goals (Dryzek 1997).

From the popular democratic perspective, participation "makes people more aware of the linkages between public and private interests, helps them develop a sense of justice, and is a critical part of the process of developing a sense of community" (Laird 1993, 345). In environmental policymaking, the popular perspective has focused attention on the role of communities in environmental protection, spawning, for example, the U.S. Environmental Protection Agency's (EPA's) efforts at "community-based environmental protection" and the National Environmental Justice Advisory Committee's *Model Plan for Public Participation* (NEJAC 1996).

What Society Expects from Public Participation

Although one can chart a historical progression from managerialism to pluralism to popular democracy, all three perspectives continue to compete in contemporary debates about how environmental policy should be made and implemented (Reich 1985).

What have evolved are society's expectations about what public participation should accomplish. Throughout the managerial era, the main justification for public involvement was accountability: to ensure that government agencies were acting in the public interest. With the ascendancy of pluralism and popular democracy, participation has been seen as a necessity for establishing what that public interest really is. The purpose of participation has shifted from merely providing accountability to developing the substance of policy.

The change in expectations is apparent in the proliferation of public participation activities in federal environmental programs in the 1990s. Federal departments and agencies have introduced much more complex forms of public participation than the traditional approaches of public comment and public hearings. The rise in the use of consensus-based participation (in which interest groups are expected to negotiate and agree on policy outcomes) has been particularly noticeable. The U.S. Department of Energy (DOE) has used consensus-based advisory committees in making cleanup decisions at its contaminated facilities around the country (Bradbury and Branch 1999). EPA's Project XL and Common Sense Initiative programs have relied on consensus-based committees (Yosie and Herbst 1998). Consensus-based grassroots stakeholder councils have sprung up around the country to decide how to manage natural resources, and participation has become integral to resource management at the Department of the Interior and the Forest Service (Weber 2000). Reflecting the ascendancy of stakeholder negotiation in developing policy, in 1990 Congress passed the Negotiated Rulemaking Act, which allows agencies to use formal negotiations among interest groups to develop proposed regulations.

One reason that participation has become more central to environmental decisionmaking is an expectation that it can temper the confrontational politics that typify environmental policy. Modern environmental policymaking has been described as beset by a group of "wicked problems," meaning "problems with no solutions, only temporary and imperfect resolutions" for which there are no "narrowly defined technical definitions and solutions" and no "clear-cut criteria to judge their resolution" (Fischer 1993, 172–173). In short, such problems are ill-suited to a managerial approach and rife with the politics that participation can address.

One example of Fischer's wicked problems is the siting of hazardous waste facilities. Greater participation in decisionmaking has been viewed as a possible remedy for NIMBY (not-in-my-backyard) syndrome, which has bedeviled governments and industries seeking to build industrial facilities across the country (Rabe 1994). Another example is the massive cleanup of cold war–era nuclear weapons facilities, in which DOE has instituted participation to stem a crisis in public trust and confidence that is undermining the department's ability to perform complex environmental management and cleanup operations (SEAB 1993). In these and many other examples, public participation is being used not only to keep government accountable but also to help agencies make good decisions, help resolve long-standing problems of conflict and mistrust, and build capacity for solving the wicked problems of the future.

What Is Public Participation?

The public participates in society's decisions in countless ways, from voting to violence and from "letters to the editor" to lawsuits. Only some of these ways are covered in this book. We define *public participation* as any of several "mechanisms" intentionally instituted to involve the lay public or their representatives in administrative decisionmaking. Such mechanisms range from town meetings at which citizens express their opinions to formally mediated negotiations in which parties write regulations; they also include advisory committees, citizen juries, and focus groups.

Our definition excludes some methods of participation that are important in their own right and have extensive traditions as well as legal foundations. For example, we exclude voting for elected officials, referenda, and initiatives as well as lobbying and citizen lawsuits. We also exclude less regulated methods (such as striking and picketing) and extralegal ones (such as violence). We focus on organized bureaucratic processes, not individual actions or power politics.

Although our definition is narrower than it could be, it is broader than some. In particular, many analysts and practitioners distinguish between *public participation* and *stakeholder involvement*. The former term generally connotes a popular democratic notion of lay citizens' involvement in local issues and the latter term a more pluralist notion of interest group involvement in policy-level questions. We make no such distinction here and use *public participation* as an umbrella term that encompasses diverse definitions of who the public is, how the public is represented, why the public is involved, and what the public is involved in.

Evaluating Public Participation: Overview

We evaluate public participation on the basis of contemporary claims about what it can accomplish in environmental decisionmaking. These claims can be summarized as five "social goals" for public participation (Beierle 1999):

Goal 1: Incorporating public values into decisions
Goal 2: Improving the substantive quality of decisions
Goal 3: Resolving conflict among competing interests
Goal 4: Building trust in institutions
Goal 5: Educating and informing the public

In this book, *success* is defined as the extent to which public participation efforts achieve these social goals. To evaluate the success of public participation as practiced in the United States over the past 30 years, we turn to the case study record and synthesize data from several hundred published studies covering 239 cases of public involvement in environmental decisionmaking. (The process of selecting cases is described in Appendix A, and the list of cases is presented in Appendix E.) Case studies were published in journals, books, dissertations, conference proceedings, and government reports. They cover diverse planning, management, and implementation activities carried out by citizens and agencies at many levels of government. Each case involves a participation process specifically designed to engage people outside of government in helping to make decisions concerning the environment (e.g., public hearings, advisory committees, negotiations, or mediation).

In preparation for our study, we coded each public participation case based on a standard conceptual model composed of more than 100 attributes that describe the context, process, and results of the particular case. This conceptual model, the social goals used to evaluate the results of each case, and the methods used to select and analyze the case studies are described in Chapter 2.

In Chapter 3, we explain how the cases measure up against the social goals and illustrate the results with specific examples. The view of public participation derived from the analysis is quite positive. In most of the cases, public values and knowledge made important contributions to the quality of decisions; the participation processes often also resolved conflict, increased trust, and educated the participants. We discuss important caveats about how positive results should be interpreted.

Next, we examine the roles of context and process. In Chapter 4, we discuss how the type of issue being debated, the preexisting relationships among parties, and the institutional context relate to success. In Chapter 5, we evaluate the relative success of different participatory mechanisms (e.g., public hearings and advisory committees) and specific process elements (e.g., the participants' degree of motivation and the quality of their deliberation). The analysis demonstrates that the process of participation, rather than its context, is largely responsible for the success or failure of public participation.

We examine the relationship between public participation and implementation in Chapter 6. The cases provide only partial support for the claim that good public participation helps smooth the path to implementation. They reveal several forces acting on implementation that are not related—or only tangentially related—to public participation.

In Chapter 7, we pull the lessons from the report together into a discussion of how to design successful public participation processes. Rather than advocating the use of particular mechanisms, we present a process of strategic design, starting with an understanding of whether public participation is appropriate, moving on to process goals, and finishing with specific features of process design.

The book concludes with a review of our main points and some suggested areas for further research in Chapter 8.

Chapter 2

Conceptual Framework and Methodology

The cases examined in this study ranged from one-day public meetings to intensive negotiations. Issues under debate included hazardous waste cleanup, facility siting, water supply, and a broad array of other environmental topics. Relationships preceding participation ranged from cooperative to combustible. Interactions were civil in some cases and circuslike in others.

In this chapter, we describe the methodology used to investigate this heterogeneous group of cases: the conceptual model used to organize each case study for analysis, the approach to evaluating those studies, and the research methods for collecting and analyzing the data.

Conceptual Model of Public Participation

To analyze a large number of cases of public participation, it is necessary to develop a conceptual model that is general enough to fit the wide variety of public participation efforts but detailed enough to identify important differences among them. The basic skeleton of our analysis has three major components: context, process, and results. Each case occurred in a particular context, used a particular process, and produced a particular set of results.

Within these three categories, we use several component attributes to describe the cases (Figure 2-1). To make sense of more than 100 pieces of data on each case, we focus attention on only the most important attributes, many of which are aggregates of multiple data points. (The disaggregated information collected for each case is described in Appendix B, along with details on the aggregation process.) The identification of important attributes draws heavily on our pilot project (Beierle and Konisky 1999).

Context	Process	Results
Type of Issue	**Type of Mechanism**	**Output**
• Policy level vs. site specific	• Selection of participants	**Relationships**
• Pollution vs. natural resource	• Type of participant	• Among public
• Topical category	• Type of output	• Between public and agency
Preexisting Relationships	• Use of consensus	**Capacity Building**
• Conflict among public	**Variable Process Features**	
• Mistrust of government	• Responsiveness of the lead agency	
Institutional Setting	• Motivation of participants	
• Level of government	• Quality of deliberation	
• Identity of lead agency	• Degree of public control	
• Lead agency's level of involvement		

FIGURE 2-1. Conceptual Model of Public Participation: Categories and Attributes

Context

This category refers to all the features of a given situation that a public participation process confronts. For example, consider some aspects that form the context of a case concerning the cleanup of the Tucson International Airport Superfund site (Di Santo 1998):

- Groundwater was contaminated from decades of industrial dumping by military contractors.
- The amount, movement, and health effects of contamination as well as the cleanup methods were quite technical and not well understood.
- Residents' complaints and lawsuits over drinking-water wells began in the 1950s; the site was put on the Superfund National Priorities List (NPL) in 1983, and local residents won an $84.5 million settlement in 1991.
- Local area residents are mostly poor, and many are Spanish-speaking.
- Migration of contaminants threatened Tucson's entire water supply.
- The U.S. Environmental Protection Agency (EPA), U.S. Department of Defense (DOD), state agencies, and local government had overlapping authority.

Understood broadly, *context* could include an infinite number of such attributes. We focus on three of the most salient: the type of issue, the preexisting relationships among members of the public and between the public and the lead agency, and the institutional setting.

The type of issue under discussion could be developing a plan to protect a threatened salmon population in Maine (Opperman 1998a) or designing strategies for meeting power needs in Tennessee (Ford 1986). Some cases involved debate over a particular site or geographic feature. In facility siting cases, for example, members of the public discussed the merits and locations of incinerators, power plants, dams, dikes, canals, mines, highways, landfills, ferry terminals, and radioactive waste facilities. Other cases required overarching policy decisions. For example, participants negotiated the fine points of regulatory policy concerning asbestos, water pollutants, industrial equipment leaks, bear hunting, worker safety, transportation, gasoline additives, wood-burning stoves, and hog farming. Some cases concerned pollution, whereas others were about the conservation and use of natural resources. Issue types also can be categorized by topic, such as the facility siting and regulatory development cases above, but also hazardous waste cleanup, permitting, and others.

Beyond the characteristics of the particular issue, we examine the quality of the preexisting relationships among the people, organizations, and institutions that addressed the issue. For example, park management at Devil's Lake State Park, where relations were good among stakeholders and economic incentives harmonized with environmental conservation (Cavaye 1997), is different from that at Promised Land State Park, where users had directly opposing interests from (and historically clashed with) state agencies (Gray and Purdy 1992).

Finally, institutional context is important. The cases took place at different levels of government and under different agencies. The degree of lead agency involvement also differed from case to case.

Process

If *context* encompasses all the conditions a public participation effort faces, then *process* is what actually happens. Various features of process may affect results. Most important is the kind of process, whether public meeting, advisory committe, or negotiation.

The paradigms of public participation are town meetings and public hearings. In considering a permit for a limestone mine on Laurel Mountain, for example, the West Virginia Division of Environmental Protection held a public meeting at which local residents could voice their concerns (Steelman and Carmin 1998). The National Park Service greatly elaborated such an approach to develop management plans for Yosemite Park by holding six years of public meetings supplemented with surveys, workshops, and mailed workbooks (Buck and Stone 1981).

In cases that require more intensive discussions or problem-solving than is possible with open meetings, advisory committees are the most common alternative. Some advisory committees have little authority and just try to keep the community informed about an issue. In the Tucson International Airport Superfund site case, for example, a largely powerless citizens' advisory board was overwhelmed by DOD, EPA, powerful companies (e.g., Hughes Aircraft and Raytheon), and the technical complexity of the remediation options

(Di Santo 1998). However, other advisory committees have considerable influence on agency decisions. In Maryland, the Department of Natural Resources (DNR) instituted a stakeholder steering committee to develop criteria for siting new power plants in the coastal area. With most stakeholders represented on the committee, DNR generally accepted the committee's advice and recommendations (McConnon 1986). Variations on advisory committees take many interesting forms. For example, "citizens juries" are mock trials in which experts testify about various policy options and then citizens, acting as the jury, develop policy recommendations (Crosby, Kelly, and Schaefer 1986).

Some processes give participants even more authority than an advisory role and involve the formal negotiation of agreements among parties. In one case, EPA conducted a negotiation on the regulations governing wood-burning stove emissions. Sixteen participants, professionally representing industry, environmental groups, and state agencies, came to an agreement that was the blueprint for EPA's regulations (Ozawa 1991b; Funk 1987). In less formal settings, participants in watershed management committees work out among themselves how to manage a common resource.

We use the term *mechanism* to refer to categories of processes (e.g., public meetings, advisory committees, and negotiations). The choice of mechanism usually determines several other characteristics about the process, such as how participants are selected, the type of people who participate, what kind of output the participants will produce, and whether the participants will seek consensus.

Beyond differences in mechanisms, other process features vary from case to case. We focus on four of these features: the responsiveness of the lead agency, the motivation of the participants, the quality of deliberation among participants, and the degree of public control over the process. These features stood out in importance in our pilot project, as they do in the broader public participation literature (Arnstein 1969; Beierle and Konisky 1999; Fiorino 1990; Gurtner-Zimmermann 1996; Hartley 1999; Innes 1998; Renn, Webler, and Wiedemann 1995; Slovic 1993).

Some of the process aspects that we discuss have their roots in the context of decisionmaking (e.g., personal qualities that help motivate participants exist before a process is initiated); however, choices about process design help determine who will participate and, therefore, what characteristics they will bring to the process. The planning, design, and execution of the process make these personal qualities relevant to a particular decision.

Results

Context and process combine to produce a set of results. We define the term *results* broadly. Of course, it includes the specific output of the public participation process, which varies from case to case. For example, residents of Boulder, Colorado, sat down in their living rooms with members of a city task force to comment on what they wanted to see in the city's master transportation plan (Kathlene and Martin 1991). The Northern States Power Advisory Task Force provided recommendations to a Minnesota state agency regarding the siting of a coal-fired power plant (Ducsik and Austin 1986). A roundtable of interests in the Chesapeake Bay reached a consensus agreement on an oyster management plan (Arnold 1996). These comments, recommendations, agreements, and other types of outputs are the results that participants intend to produce, but they are not the only results of public participation.

The act of participating may improve relationships among interest groups or foster trust in the lead agency. The once-burning Cuyahoga River in Ohio has become the subject of cooperative ventures among competing interests (Becker 1996). In contrast, conflict over the U.S. Army Corps of Engineers' plans for a barge canal near New Orleans boiled over in public meetings, which degenerated into "a parade of demonstrators marching around the room" (Mazmanian and Nienaber 1979, 92).

We also consider results in terms of capacity building, where *capacity* refers to the public's ability to understand environmental problems, get involved in decisionmaking, and act collectively to implement change. We examine one aspect of capacity building here: increasing the public's knowledge and understanding of environmental issues. Capacity building has been examined elsewhere in terms of the formation of relationships and the development of formal institutions (Beierle and Konisky 2001).

Future Development of the Conceptual Model

The conceptual model we use to analyze public participation in this book contains what we believe to be the most important elements of most public participation processes, but it is not comprehensive. The database of cases contains a larger set of elements pertaining to the context, process, and results categories that future research will address. These elements include measures for the complexity of decisions, the degree of scientific uncertainty, the geographic scale of the issue, the complexity of jurisdictional authority, the degree of programmatic authority, and a broader set of measures of capacity building.

Evaluation Framework

We use the five social goals listed in Chapter 1 to evaluate the results category of the conceptual model (Figure 2-2). We evaluate output in terms of the extent to which public values were incorporated into decisions and whether the substantive quality of decisions was improved. We evaluate relationships

Output

- Incorporating public values into decisions
- Improving the substantive quality of decisions

Relationships

- Resolving conflict among competing interests
- Building trust in institutions

Capacity Building

- Educating and informing the public

FIGURE 2-2. Social Goals Evaluated in the "Results" Category of the Conceptual Model

in terms of the extent to which conflict was resolved among competing interests and trust was built in the lead agency. We evaluate capacity building in terms of whether the public became better educated and informed about environmental issues. We introduce each social goal here and then describe them in greater detail in Chapter 3.

Goal 1: Incorporating Public Values into Decisions

This goal is fundamental to democracy and has been the driving force behind challenges to the managerial model, which traditionally has recognized a rather limited set of public values. The risk perception and communication literature, for example, outlines dramatic differences in the way that the lay public and experts view risk (Slovic 1992; Stern and Fineberg 1996). Whereas experts generally define risk in terms of the value that the public places on maintaining good health, the public brings a much more complex set of values to their understanding of risk. The public, though, is not monolithic; members of the public can have widely differing values that affect their views about environmental issues (Bauer and Randolph 1999).

We evaluate the extent to which participants come to agree on values-related issues and how much influence public values have on policy decisions. Because the public holds a range of views, it is particularly important to identify who is represented in the public participation process.

Goal 2: Improving the Substantive Quality of Decisions

The public is widely recognized as a source of knowledge and ideas for making decisions (Fiorino 1990; Raffensperger 1998; Stern and Fineberg 1996). The public may improve the substantive quality of decisions in several ways, such as by offering local or site-specific knowledge, discovering mistakes, or generating alternative solutions that satisfy a wider range of interests.

Goal 3: Resolving Conflict among Competing Interests

The environmental regulatory system in the United States was born of conflict between environmental and industrial interests, and conflict has persisted as the system has matured. As many observers have recognized, substantial amounts of money and energy have been consumed by court battles and other kinds of conflict while environmental problems remain unresolved.

The goal of conflict resolution is based on the argument that collaborative rather than adversarial decisionmaking is more likely to result in lasting and more satisfying decisions, potentially averting the litigation and gridlock that characterize much environmental decisionmaking (Susskind and Cruikshank 1987).

Goal 4: Building Trust in Institutions

Many analysts have highlighted the dramatic decline in public trust of government over the past 30 years (PRC 1998; Ruckelshaus 1996). A healthy dose of skepticism is important for ensuring government accountability. However, the decline is also symptomatic of what some claim to be a general decline in the networks, norms, and trust that bind civil society together (Putnam 1995).

As trust in the institutions responsible for solving complex environmental problems decreases, their ability to resolve those problems is seriously circumscribed. Research suggests that one of the few ways agencies can try to rebuild trust is through allowing greater public involvement and influence in decisionmaking (Schneider, Teske, and Marschall 1997; Slovic 1993).

Goal 5: Educating and Informing the Public

Increasing public understanding of environmental problems builds capacity for solving those problems. Education in this context refers to more than science lessons. It integrates information about the problem at hand with participants' intuition, experience, and local knowledge to develop a shared understanding and a collective perception of solutions. Such an education helps the public build the capacity needed to formulate alternatives and helps to level the playing field between the public and government (Stern and Fineberg 1996).

Defining Success

We use performance on the five social goals to define *success*, but the five goals are not necessarily related. In some cases, goals may compete. Resolving conflict may require compromises that sacrifice decision quality. Focusing on extensive public education may be perceived as manipulative, leading to a loss in trust. In other cases, goals may be complementary. The question is largely empirical, and we show in Chapter 3 that the goals are, in fact, related.

Other Evaluations of Public Participation

In 1983, Rosener (1983, 45) described the obstacles to evaluation: "The participation concept is value laden; there are no widely held criteria for judging success and failure; there are no agreed-upon evaluation methods; and there are few reliable measurement tools." Analysts of public participation have since risen to the challenge and have produced a rich and varied body of evaluation studies. Yet no consensus has developed on the "right" approach to evaluation, largely because of different views about the purpose of public participation.

One set of analysts has focused on public participation as a way to democratize bureaucratic decisionmaking. Their evaluations examine the process of participation—comparing actual cases with an ideal model. For example, Renn, Webler, and Wiedemann (1995) draw on Habermas's theory of communicative action to develop an evaluation framework based on "fairness" and "competence" in public participation processes. Fiorino (1990) derives evaluation criteria from democratic theory, arguing that participation should involve face-to-face discussion in which citizens share in decisionmaking and participate on an equal footing with experts and agency officials. Other evaluators rely less on theory and more on rules of thumb about which procedural attributes have been consistently successful over time (Ashford 1984; Blahna and Yonts-Shepard 1989; Crosby, Kelly, and Schaefer 1986; Peelle and others 1996).

A second set of analysts has viewed public participation more strategically—as a way for particular interests to achieve their goals. They generally take the perspective of one set of interests in a decision, such as a community group or public agency. For example, Arnstein (1969) founded the field of public participation evaluation with a "ladder of public participation," which measures the degree of power that citizens have over decisionmaking. Other evaluations have taken the perspective of a lead agency. In a survey of 22 evaluations of public participation programs, Sewell, Phillips, and Phillips (1979) report that evaluations performed by agency personnel tend to measure success in terms of the degree of public acceptance of agency programs and improved agency image. Although public acceptance and an improved image may be an agency's primary goals, other participants in the process are likely to have very different goals.

A third set of analysts has taken a perspective similar to that used in this report, that is, viewing public participation as a means to achieving broad social goals. For example, in the first comprehensive analysis of mediated environmental disputes, Bingham (1986) analyzes whether

continued on next page

Other Evaluations of Public Participation—*continued*

mediation resolves disputes and leads to agreements that can be implemented. In a similar vein, Coglianese (1997) evaluates federal regulatory negotiations on the basis of whether they reduce litigation and save time. Leach, Pelkey, and Sabatier (2000) examine watershed groups to determine whether they (among other things) foster education and outreach, build capacity, resolve conflict, and lead to environmental improvements.

Finding a definitive answer to the question of what is the "right" way to evaluate public participation is neither likely nor desirable. Each approach to evaluation poses—and hopefully answers—interesting questions that collectively inform our understanding of this complex social process.

Intentionally absent from the five social goals is implementation. Many people regard changes in the real world—achieved by implementing decisions—to be the primary social goal against which public participation should be evaluated. However, the implementation of decisions depends on many factors other than public participation; funding and regulatory authority are two. The time scale of implementation is often much longer than that of participation as well. Ultimately, we would like to understand whether good public participation (as measured by the five social goals) can help the implementation process; we examine this relationship in Chapter 6.

Methodology

The foundation for our research is a database, constructed according to the conceptual model and evaluation framework, that houses extensive information on 239 cases of public participation in environmental decisionmaking undertaken in the United States over the past 30 years. We derived data on the cases using a "case survey" methodology (Bullock and Tubbs 1987; Larrson 1993; Lucas 1974; Yin and Heald 1975). A case survey is analogous to a normal closed-end survey, except that a reader-analyst "asks" a standard set of questions of a written case study rather than of a person. It is a formal process for systematically coding relevant data for quantitative analysis from many qualitative sources. The case survey process is explained in detail in Appendix A, but it is important to touch on some of its basic features here.

Case Selection and Coding

We screened more than 1,800 case studies—drawn from journals, books, dissertations, conference proceedings, and government reports—and ultimately identified those that detail the 239 cases in the data set. The screening criteria were selected to ensure that each chosen case study had sufficient information on the context, process, and results of a case that involved

- public participation in environmental decisionmaking that occurred in the United States over the past 30 years;
- a discrete mechanism (or set of mechanisms) intentionally instituted to engage the public in administrative environmental decisionmaking, such as public hearings, advisory committees, or environmental mediation;
- the participation of nongovernmental citizens;
- the involvement of at least one public perspective (such as an environmental group or a community group) other than the regulated community; and
- either an identifiable lead agency or an agency for which the output of the process would be immediately relevant.

The authors and a research associate coded each case for more than 100 attributes that included aspects of the context of participation, the process of participation, and the outcomes achieved (see Tables B-1 to B-4 in Appendix B). Each attribute was assigned a score—usually low, medium, or high—on the basis of standardized descriptors of what the score meant for that attribute. The attribute scores were given one of three weight-of-evidence indices, ranging from "solid evidence" to "best informed guess." (Data with the lowest weight of evidence were not used in our analysis.) The scores were accompanied by a written entry that described the attribute in qualitative terms.

Each case was coded by one or two researchers. To ensure consistent coding, a process of intercoder reliability testing and training was used. Pairs of researchers read and coded the same subset of randomly selected case studies independently, then compared codes. Standard practice requires that coders consistently achieve two-thirds agreement (Larsson 1993). The intercoder reliability test was repeated periodically throughout the coding process. Approximately 10% of the cases were used in the testing.

Data Analysis

After coding all the cases, we proceeded to the data analysis phase of the project. Data analysis consisted of summary counts of scores, an informal review of the qualitative information accompanying the scores, and comparative statistical techniques (described in Appendix C). In the main body of this book, we rely as much as possible on figures to make our principal points. However, some statistics are used in the discussion, so a few notes are warranted here.

We used two statistical techniques. *Bivariate correlation* examines the relationship between two variables. A correlation of 0 means that there is no relationship between the variables, and a correlation of 1 means that the variables are perfectly correlated. The type of ordinal data used here (e.g., low, medium, and high scores) required the use of a nonparametric correlation technique. Calculating correlation coefficients involved the use of contingency tables—essentially, cross-tabulations of the results for two attributes—and counts of the numbers of matching and nonmatching pairs of results. The social science literature has no fixed standard for what level of correlation should be regarded as high or low. We consider correlations above 0.45 as "high," correlations between 0.3 and 0.45 as "moderate," and correlations below 0.3 as "low."

Multivariate ordered probit regression explains the variation found in a dependent variable based on the variation in a series of independent variables. The main advantage of multivariate analysis over bivariate correlation is the ability to control for the influence of one (or more) variables when examining another. In the multivariate analysis, we assume that the social goals are the dependent variables and that the degree to which they are achieved depends on the context and process attributes, which we regard as independent variables. For all the statistical analyses, a "significant" relationship means that its associated probability is greater than 95% (i.e., $p < 0.05$).

Caveats and Analysis of Bias

Even though it has been used in the policy analysis and business literature, the case survey methodology is still somewhat experimental, so a few important caveats are warranted. The quality of the data used in a case survey is only as good as the quality of the case studies that provide the data. In coding the cases, we sought to distinguish, to the extent possible, the opinions of case study authors from the facts reported. Inevitably, insights into each case depended heavily on the case study authors' interpretations of events. Case study authors focused on different facets of each case, which meant that we could not code every attribute for every case. Our discussion of any given attribute typically refers to 100–200 cases, and the particular set of cases varies from attribute to attribute.

In some instances, more than one case study described the same event. Such multiple accounts were useful for cross-checking the facts of the case and also enabled coding with greater confidence. Occasionally, discrepancies emerged, and we used our judgement to select among competing descriptions. When the correct choice was not clear, the attribute either was not coded or was coded with the lowest level of confidence. In either case, it was not part of the data analyzed for this book.

Like any meta-analytical technique, the benefit of covering a breadth of cases comes at the expense of depth in any given case. In deriving broad lessons across cases, the idiosyncrasies and nuances of individual cases can be obscured. Complex causalities also are difficult to handle with standard statistical techniques. In our analysis, we assume that the goals are influenced by context and process attributes. In some cases, causality may work in the other direction.

A final important issue for case surveys is whether the data set is biased. Given the high number of successful cases, we might be particularly concerned about a bias toward success. We could not check for bias by going back and comparing the 239 cases to a randomly sampled control group, but we could examine possible sources of bias quantitatively and qualitatively. The examination of bias in Appendix D is summarized here.

Conducting a case survey entails completing various steps before analyzing the data, and bias may be introduced at each step. For our study, the six steps were as follows.

1. An agency chooses to use a public participation process.
2. A case study author decides to write about a particular public participation process.
3. A case study author chooses study methods, data interpretation methods, and the emphasis and tone of the narrative.
4. We identify case studies in our literature review.
5. We use a screening process to select case studies for analysis.
6. We code the attributes of each case.

We conclude that the potential for success bias from Step 1 and from Steps 4–6 is small. Steps 2 and 3 are more likely to be sources of success bias; for example, authors may prefer to write about successful cases, or they may overemphasize the positive aspects of the case.

Two considerations temper this concern. First, case study authors do not necessarily have an incentive to write about only successful cases. Many of our cases come from doctoral dissertations or other studies in which multiple cases are compared and unsuccessful cases provide as much insight as—or perhaps more insight than—successful ones. Second, authors define the success of stakeholder processes differently. Some authors consider success to be participants reaching consensus, others consider success to be a fair process, and still others consider success to be progress toward implementation. Very few authors use the same evaluation criteria that we do. Even if a success bias exists, its influence on our results will be significantly tempered by the wide range of definitions of *success*.

In Appendix D, we examine quantitatively whether our conclusions are affected by case study authors' biases in picking cases or writing about them. The principal conclusion from our analysis is that although a moderate bias toward successful cases may exist, it has little impact on our main conclusions.

Chapter 3

The Social Goals
of Public Participation

The Great Lakes have long been a symbol of the worst environmental pollution and the best responses to it. In the 1960s, Lake Erie was declared dead—choked with algae and starved of oxygen. Then the Cuyahoga River, which feeds the lakes, caught fire in the summer of 1969.

However, over the past 30 years, water quality in the lakes has improved dramatically through the cooperation of state, provincial, and federal agencies with citizens and local authorities. In 1972, the United States and Canada signaled their joint resolve to tackle water pollution by signing the Great Lakes Water Quality Agreement. Fifteen years later, the International Joint Commission (IJC), which oversees the agreement, approved an amendment that targeted the most vexing remaining insults to the Great Lakes: 43 highly contaminated rivers and bays in which pollution restricted swimming, fishing, and other uses. Residual contamination from these areas threatened the larger lake system. In what is known as the Remedial Action Plan (RAP) process, the IJC called on state and provincial agencies to join with the public to develop and implement plans for cleaning up these areas.

The RAP projects provide an interesting lens with which to view public participation in action (Beierle and Konisky 1999). Two cases illustrate the different outcomes public participation processes can produce and how they can be evaluated by using social goals.

The first case began in 1987 in Buffalo, New York, when the Buffalo River Citizens Committee (BRCC) formed to advise the New York Department of Environmental Conservation (DEC) on restoring the lower reach of the Buffalo River, where it enters into Lake Erie. Participants in the BRCC included environmental advocates as well as representatives of business, local government, labor unions, and other community organizations.

Members of the committee worked for two years to devise a cleanup plan. Many members had a great deal of scientific knowledge and training, and they became a valuable technical resource for the DEC (Kellogg 1993b). By the time the DEC submitted the cleanup plan to the IJC for approval, members of the advisory committee reported that their experience together had helped to resolve conflict and build trust around a shared commitment to the river. The IJC agreed with the members of the BRCC that the public involvement effort had been a great success.

The second case took place 200 miles to the west, where a similar RAP process was under way in Detroit, Michigan. After its formation in 1986, the Binational Public Advisory Council (BPAC) began work to advise the Michigan Department of Natural Resources (DNR) on a plan to restore the Detroit River, a 30-mile channel that connects Lake Huron (via Lake St. Clair) with Lake Erie. Members were citizens at large and representatives of many interests groups.

Many participants in the BPAC reported that the state agency cared little for their input. "Over time attendance at BPAC meetings ... dwindled to just a few people and public participation consisted mostly of listening to what government agencies had to say" (Becker 1996, 476). Amid a high level of distrust and contentiousness within the BPAC, DNR submitted the Detroit River RAP plan to the IJC in 1991. By that point, the agency had "lost all trust of everybody," according to one participant (Becker 1996, 467). In a subsequent meeting about implementing the RAP plan, half the committee—its environmental, labor, and citizen representatives—walked out, abandoning the process. The IJC review of the Detroit River process was the most critical of any of the RAP reviews (Becker 1996).

By most standards, the Buffalo River case stands out as a success and the Detroit River case as a failure. However, the social goals introduced in Chapter 2 provide a consistent and common framework for evaluating what succeeded and what failed in these processes. In the Buffalo River case, participants influenced the cleanup plan through the diverse values they brought to the table and their extensive knowledge of the river. They resolved conflict among themselves, developed trust in the state agency, and learned a great deal about water quality. In the Detroit River case, participants learned a great deal about water quality, but they also learned that the state agency had little interest in their input to the plan. Their frustrations deepened mistrust, and conflict among participants worsened.

In this chapter, we report on the evaluation of all 239 cases using the social goals framework. We discuss each of the social goals individually, then conclude with the calculation of an overall aggregate measure of success for each case. In later chapters, we investigate why some cases achieve the social goals and some do not.

Goal 1: Incorporating Public Values into Decisions

This goal measures the extent to which participants influence policy decisions. Very few cases are mostly about technical decisions; nearly every one engages public values in some way. We assume, then, that any influence participants have on outcomes reflects the values of the participants. In scoring this goal, we made no judgement about whether public influence made decisions better or worse; we were interested only in how much public values affected the final decisions made or ratified by lead agencies.

The majority of cases scored high on this goal. Of 195 cases, 58% received high scores, 27% received medium scores, and 15% received low scores (Figure 3-1). In the cases that received high scores, public input created or substantially changed decisions. In the cases that received medium scores, public input may have informed the analysis but did not substantially affect the decisions made. In the cases that received low scores, public input had little impact on analysis or decisionmaking.

The development of the Louisiana Black Bear Conservation Plan is one example in which public values had a large impact on decisionmaking. A stakeholder group representing diverse values and the full range of environmental and economic interests in the area (the Black Bear Conservation Com-

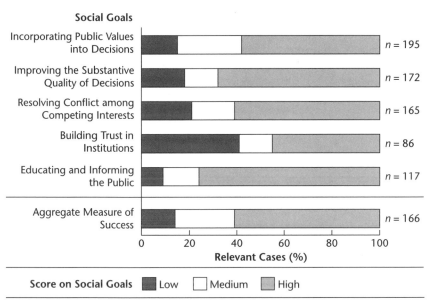

FIGURE 3-1. Social Goals of Public Participation

Note: n = total number of cases scored.

mittee) developed a plan to restore black bear populations in Louisiana after the species was proposed for listing under the Endangered Species Act. The committee devised a plan that effectively balanced economic and environmental values by protecting the bear while limiting onerous permitting requirements for timber and agricultural interests (Merrick 1998a). The committee's plan became the basis for the Fish and Wildlife Service's official recovery plan.

Other cases took a more explicit approach to identifying public values and incorporating them into decisions. For example, the Maryland Department of Natural Resources convened a stakeholder committee when the agency needed to select the optimal coastal zone sites for energy developments. The committee identified siting criteria that reflected public values, such as the protection of prime agricultural land, wetlands, and other natural areas. The agency then combined the values-based criteria with standard technical criteria to identify potential development areas (McConnon 1986).

In the few cases that received low scores, public values had no influence on decisionmaking. In some of these cases, participants could not reconcile their diverse "public values" and agree on a course of action. In other cases, agencies were not responsive to public input. For example, in western Kern County, California, the county government approved a proposed expansion of a toxic waste dump despite a public advisory committee's strong objections (Cole 1999). The county may have had very good reasons for approving the dump, or it may not have. What matters in the scoring of the values goal is that the agency chose not to follow the suggestion of the public advisory committee.

In cases where public values do affect decisionmaking, the question that arises is "whose values?" How a small group of participants can or should represent a broader public is a question with a long history in political philosophy. Most discussions of public participation measure representation in terms of either socioeconomic characteristics or interest orientations.

In nearly 60% of the 63 case studies that were coded for socioeconomic representativeness, participants were not at all representative of the wider public. In 58% of the 74 cases with information on interest group representation, participants or case study authors identified interests that were missing from the table. The fact that participants were not necessarily a good reflection of the wider public might not be important if participants sought out input from the wider public directly. However, of 129 case studies with relevant information, participants consulted the wider public in 39% of the cases, consulted to some extent in 27% of the cases, and did not consult at all in 34% of the cases. Although participants had a high degree of influence in the majority of cases, important questions remain about how well those participants reflected the values of the public they were meant to represent.

Scoring the Social Goals

We scored the five social goals on the basis of the following descriptive information. Also included are some other questions that should be considered in assigning meaning to the social goals scores. Future evaluations can use these goals as a starting point.

Goal 1: Incorporating Public Values into Decisions

How much influence is the public having on decisions made? We defined the scores as follows:

- high—Public input made or substantially changed decisions.
- medium—Public input may have informed analysis but did not significantly affect the decisions made.
- low—Public input had little impact on analysis or decisions.

When asking questions about public values, we need to know whose values we are talking about. Are the participants socioeconomically representative of the wider public? Are all the interests at the table? Are processes in place for soliciting input from the wider public?

Goal 2: Improving the Substantive Quality of Decisions

Are stakeholders improving decisions through creative problem solving, innovative ideas, or new information? We scored this goal based on eight criteria, divided into two sets. The first set measured whether decisions were superior to likely alternatives in terms of cost-effectiveness, joint gains, the opinions of participants, or other measures. We defined the scores for this set as follows:

- high—Quality increased.
- medium—Quality did not change.
- low—Quality decreased.

The second set of criteria measured whether participants added information, provided technical analysis, contributed innovative ideas, or contributed a holistic perspective. We scored these criteria as "yes" or "no." The first and second sets of criteria were combined into a single measure of substantive quality.

Goal 3: Resolving Conflict among Competing Interests

Was conflict that was present at the beginning of the process resolved by the end? Scoring this goal required combining information about the

continued on next page

Scoring the Social Goals—*continued*

preexisting level of conflict with information about conflict at the end of the process. We defined the scores as follows:

- high—Preexisting conflict was resolved, or good relationships were maintained.
- medium—Conflict was resolved only on some issues or only among some participants.
- low—Preexisting conflict was not resolved, or conflict was made worse.

To interpret the significance of the conflict resolution score (but not to influence the score itself), we asked two more questions, which helped us understand whether conflict was resolved or simply avoided: Was conflict avoided by avoiding contentious issues? Was conflict avoided because certain parties were excluded or chose not to participate?

Goal 4: Building Trust in Institutions

Was mistrust of agencies that was present at the beginning of the process lessened by the end? Like resolving conflict, this goal required combining information about preexisting trust with information about the level of trust at the end of the process. We defined the scores as follows:

- high—Trust was built by the process, or a state of high trust was maintained.
- medium—Trust was improved only moderately or only among some participants.
- low—Trust decreased, or a state of low trust was not improved.

Many instances of declining trust stem from society-wide mistrust of institutions. It is crucial, then, to determine how broadly trust formation extends beyond participants to the wider public.

Goal 5: Educating and Informing the Public

Did the public learn enough about the issue to actively engage in decisionmaking? We defined the scores as follows:

- high—Participants learned a great deal about the issue under debate, enabling them to be effective partners in decisionmaking.
- medium—Participants learned about the issue, but not enough to feel effective in the process.
- low—Participants learned little about the issue.

The importance of educating and informing the public, like that of building trust, extends beyond the participants. We should ask, then, about the extent and effectiveness of educational outreach to the wider public.

Goal 2: Improving the Substantive Quality of Decisions

Public participation should not only help incorporate public values into decisions but also improve the substantive quality of decisions, as measured by relatively uncontroversial quality criteria. Are stakeholders making decisions that are more cost-effective or more satisfying to a range of interests? Are they improving the information or analytical foundations on which decisions are made? At the very least, we hope that the price of including public values is not paid by lowering these other dimensions of the quality of decisions.

One of the emerging challenges to the growing role of public participation is concern that the public makes bad decisions. The argument has been framed primarily in terms of insufficient use of scientific information and technical analysis. A prominent report issued in the late 1990s suggested that scientists and scientific information are not well integrated into most public participation processes (Yosie and Herbst 1998). EPA's Science Advisory Board took up the issue and produced a report in August 2001, concluding that absent "substantial financial resources, adequate time, and high-quality staff … stakeholder decision processes … frequently do not do an adequate job of addressing and dealing with relevant science" (U.S. EPA SAB 2001, 3).

However, the case studies suggest that the public is perfectly capable of improving decision quality. Of the 172 cases scored for this goal, roughly twice as many cases received high scores (68%) as those that received medium scores (14%) and low scores (18%) combined (Figure 3-1). The measure of substantive quality is an average of scores on the following eight separate but related quality criteria (described more fully in Appendix B and in Beierle 2000):

- cost-effectiveness—Do the decisions or recommendations made by participants lead to actions that are more or less cost-effective than a probable alternative in solving an environmental problem?
- joint gains—As a result of the agreement, are some participants better off without any participants being worse off?
- opinion—Do participants or case study authors feel that decisions are better than a probable alternative, not according to concrete criteria but in terms of general satisfaction with an outcome or in terms of a range of quality criteria?
- added information—Do participants add information to the analysis that is not otherwise available?
- technical analysis—Do participants engage in technical analyses to improve the foundations on which decisions are based?
- innovative ideas—Do participants come up with innovative ideas or creative solutions to problems?
- holistic approach—Do participants introduce a more holistic and integrated way of looking at an environmental problem?

- other measures—Do participants improve the technical quality, the environmental benefits, or other aspects of a decision?

Various criteria are appropriate because different kinds of processes affect decision quality in different ways. An agreement developed through mediation, for example, can be evaluated against a likely alternative. However, the contributions that citizens make at public meetings must be evaluated specifically, rather than by the decisions ultimately made.

Some examples illustrate how participation can improve substantive quality. Relatively few cases discuss the cost-effectiveness of stakeholder-based decisions, but the Fernald Citizens Task Force stands out as an excellent example of stakeholders developing cost-effective plans. The U.S. Department of Energy (DOE) charged the task force to advise the agency on the remediation of its nuclear weapons facility in Fernald, Ohio. DOE estimates that the resulting plan saved taxpayers more than $2 billion over the life of the project (Applegate 1998).

Participants also can improve decisionmaking by reframing issues. Narrow questions about water quality might be answered with watershed solutions; environmental cleanup decisions can become economic development plans; resource permitting debates may result in comprehensive resource management planning. For example, in a mediation regarding the damming of the Snoqualmie River in the 1970s, stakeholders refocused a narrow issue about building a dam into a more holistic set of issues that included "provid[ing] some level of flood control, ensur[ing] the continued economic viability of the farmers and the towns, and build[ing] the kind of land use plans and controls that maintain the valley as a greenbelt with broad recreational value" (Cormick and Patton 1980).

Goal 3: Resolving Conflict among Competing Interests

This goal measures the extent to which conflict that existed before the process started (or emerged during the process) is resolved. The focus is not on relationships between the public and government but on relationships among participating groups within the public.

Of 165 cases coded for resolving conflict among competing interests, 61% received high scores, 18% received medium scores, and 21% received low scores (Figure 3-1). This measure has to account for how much conflict existed at the beginning of a process as well as at the end to determine the change in conflict. Therefore, a high score indicates that a bad situation was improved or that a good situation was maintained. (Most of the high scores—80%— reflect improved situations). A medium score indicates that conflict was resolved on some issues or among some participants. A low score indicates that a bad situation was not improved or that a good situation deteriorated.

Two cases concerning the design and location of public works infrastructure illustrate the gulf between cases that received high scores and those that received low scores on resolving conflict among competing interests.

A clear success story of conflict resolution is the Central Arizona Water Control Study project, in which the U.S. Army Corps of Engineers (the Corps) was charged with designing and siting facilities for flood control and water storage around Phoenix, Arizona. The Corps initiated the study in 1978 in an atmosphere tainted by earlier fights over dam building. The community was polarized into factions for and against the dam. The process emphasized several approaches to public involvement, including a stakeholder advisory committee, a series of public meetings, and various efforts to solicit information and opinions from the wider public. Ultimately, the process produced a plan that balanced flood control and water storage needs with environmental and social concerns in a way that was acceptable to the many parties involved. The process "helped to resolve more than a decade of controversy and bitter attacks and facilitated the development of broad support for a new plan" (Brown 1984, 331).

An effort to site power lines in Minnesota had a very different outcome. Participation consisted of a series of citizen committees, public hearings, and public meetings held in affected counties. Instead of resolving conflict, the meetings only highlighted the extreme controversy between energy companies and the rural residents whose land would be bisected by the power lines. Ultimately, the two sides battled over the proposed projects in court and on the ground, resulting in nine major lawsuits, many direct action protests, acts of vandalism, and even one reported shooting (McConnon 1986).

The difference in conflict resolution between the cases in Arizona and Minnesota is stark. However, the absence of conflict at the end of a process does not necessarily signal that all is well. The contentious issue may not have been resolved but avoided—either by not discussing controversial issues or by excluding controversial participants. Of the 100 cases that received high scores on resolving conflict among competing interests, in 33 cases either participants intentionally left conflictive issues off the table or certain important parties were missing from discussions. Although we should not completely discount the reduction of conflict in these cases, we should question its social significance. The extent to which avoiding controversy may come back to harm participatory decisionmaking is taken up again in the discussion of implementation in Chapter 6.

Goal 4: Building Trust in Institutions

If the important issue among various groups within the public is conflict, then the central issue between the public and government is lack of trust

(Ruckelshaus 1996; SEAB 1993). Even though most people have an intuitive understanding of what *trust* means, it is a slippery concept to define, and various definitions coexist in the literature (Fischoff 1999; SEAB 1993). Our meaning of *trust* involves two components: competence (the ability to do what is right) and fiduciary duty (the will to do what is right) (SEAB 1993). Trust in institutions means the public believes that a lead agency is capable of understanding and serving the public interest (however defined) and obliged to do so. In coding the case studies, we sought a range of evidence regarding changes in trust, including changes in the public's perceptions about an agency's credibility, legitimacy, or competence as well as information about changes in rapport or respect between participants and a lead agency.

The results for building trust in institutions are the most polarized between high and low scores of any of the goals. Of 86 cases, 45% received high scores, 14% received medium scores, and 41% received low scores (Figure 3-1). Like resolving conflict, we measure this goal relative to the preexisting level of trust. A high score means that participants and agencies either built trust from a low-trust situation or maintained trust from a high-trust situation. (Most of the high scores—74%—reflect trust building.) A medium score means that trust improved moderately or only among some participants. A low score means that trust decreased or that a state of high mistrust did not improve.

The creation or destruction of trust is best illustrated by contrasting two cases. The Buffalo River case (described at the beginning of the chapter) illustrates how participation can build trust. For the BRCC and the New York DEC to work together to develop a cleanup plan for the lower Buffalo River, they had to overcome what one participant characterized as "a lack of trust between the agency and the committee" (Kellogg 1993b, 242). One of the outcomes of the process was a much higher degree of trust and confidence in the DEC. Interaction between the BRCC and DEC staff built respect and trust in large part because the citizens realized that the agency shared their commitment to the health of the river.

The U.S. Environmental Protection Agency's (EPA) remediation program for cleaning up the Lipari Landfill in New Jersey, one of the nation's worst hazardous waste dumps, had a very different result in terms of the relationship between the public and the lead agency. In the early phase of the project, the public participated through a series of public meetings. Citizens were so optimistic about EPA's involvement that they actually cheered agency personnel at one of the first public meetings (Kauffman 1992). Controversy over EPA's cleanup decision soon followed. The process turned into one characterized by "conflict, heated public debate, and tense relations between the USEPA and the community" (Kauffman 1992, 125). Ultimately, "residents came to believe that EPA officials did not care about their concerns," a sentiment which led to years of additional conflict and controversy (Kaminstein 1996, 461).

The two cases illustrate an interesting contrast in the potential for public participation to strain or build trust. But they highlight another important issue. In the Buffalo River case, trust was built among a relatively small group of participants. In the Lipari case, the community as a whole lost its faith in EPA. To understand the social significance of building trust, we need to know how wide the circle of affected parties extends. Unfortunately, information about building trust outside of a small group of selected participants exists in relatively few cases—only 44. Of these, more than half (52%) are cases reminiscent of Lipari; community-wide trust in agencies decreased. Only 23% received high scores, and 25% received medium scores. Because many of the cases lacked outreach efforts (and therefore, much of the public had little knowledge that these participatory efforts were even occurring), the record of building broader public trust is probably even worse than our data indicate.

Goal 5: Educating and Informing the Public

This goal was scored high most consistently across case studies. Of 117 cases, 77% received high scores, 15% received medium scores, and 9% received low scores (Figure 3-1). In the cases that received high scores, participants learned a great deal about the issues under debate—enough to be effective participants in the process. In the cases that received medium scores, participants felt that their level of knowledge hampered active involvement. In the cases that received low scores, participants learned very little.

Participants typically learn about relevant technical issues through workshops, reports written by technical advisory committees, and direct deliberation with experts. Interesting models for processes that emphasize education are citizen juries run by the Jefferson Center in Minnesota (Crosby, Kelly, and Schaefer 1986). In a citizen jury, a panel of socioeconomically representative citizens (the "jury") listens to testimony and asks questions of a series of experts (the "witnesses") and renders their informed judgement on policy topics. In processes run by the Jefferson Center, topics (e.g., electricity restructuring, feedlot regulation, traffic congestion pricing, comparative risk assessment, land use and growth management, and agricultural pollution of surface water) are often quite technically and socially complex. Nevertheless, participants consistently learned a great deal and thus were able to provide insightful policy recommendations.

Because high levels of education appear to be the norm, it is interesting to ask why some cases failed in educating and informing the public. In some cases, agencies withheld information. For example, in a study of chemical weapons disposal, "the public felt that they were left in the dark about major program decisions, such as the schedule, technology and program design.... 'They act as if we are supposed to accept and have faith that they've made all the correct decisions without telling us anything'" (Shepherd and Bowler 1997, 731). In other cases, no effort was made to assist the public in understanding information. When insufficient attention was devoted to educating and informing the public, participants remained largely powerless to engage effectively in decisionmaking.

Although most of the cases did a good job of educating the participants, they did not do well in providing educational outreach to the wider public. Of the 98 cases with relevant data, 29% received high scores on educational outreach, 22% received medium scores, and 49% received low scores. In some cases, outreach efforts simply failed, but in most cases, participants and project planners put little effort into it. For some negotiation and mediation processes, the norms of the process actively discourage outreach in an attempt to protect sensitive negotiations.

Developing an Overall Measure of Success

Across the five main goals, the public participation cases were quite successful. But were they all successful at the same things? In cases where participants resolved conflict, did they also become more trusting of agencies? Was improved education correlated with improved substantive quality? The answers to these kinds of questions tell us whether we can speak of success in unitary terms, or whether we need to focus on each goal individually.

Several conceptual reasons suggest that combining data on the five goals into a unitary measure is appropriate. That a group of traditionally adversarial stakeholders reached a consensus on a set of recommendations is evidence that they resolved conflict among competing interests, perhaps improved the substantive quality of decisions in terms of joint gains, and came at least part of the way toward incorporating public values into decisions. Similarly, one of the keys to improving the substantive quality of decisions may be adequately educating and informing the public about the issues under discussion. An agency's responsiveness to public values may build public trust.

Empirical relationships identified in the data largely uphold the conceptual relationships. The majority of the bivariate correlations among goals are high (i.e., greater than 0.45). In only 20% of the cases scored for three or more goals are high and low scores on different goals mixed.

Accordingly, we combine the five goals into a measure that we refer to throughout the rest of this book as the "aggregate measure of success." The calculation of this measure and the quantitative relationships among the social goals are explained in Appendix B. As shown in Figure 3-1, 166 cases have an aggregate goal score; 61% of these cases received high scores, 25% received medium scores, and 14% received low scores.

Interpreting Success

Considerably more public participation cases in our database produced good outcomes than produced bad outcomes. As a group, the cases were most successful in educating and informing the public and least successful in building trust in institutions. Falling in between were the results for incorporating public values into decisions, improving the substantive quality of decisions, and resolving conflict among competing interests. As an indication of the outcomes of a varied set of stakeholder processes, the case study pool gives an optimistic view of what such processes can accomplish.

Yet in thinking about the social significance of achieving the goals, it is important to remember the qualifications discussed throughout this chapter. All the qualifications concern how broadly participants and project planners draw the circle of information and influence around public participation processes. Many cases lacked significant outreach, either to inform the wider public or to incorporate their values into decisionmaking, despite the fact that the active participants often were not socioeconomically representative or did not reflect the full range of relevant interests. The tension between achieving the social goals among participants and failing to engage the wider public is apparent across the case studies; we return to this topic in later chapters.

Chapter 4

The Context of Public Participation

Environmentalists hail the 1973 Endangered Species Act (ESA) as one of the most important pieces of environmental legislation in the United States. However, many private landowners revile the law because of a fact largely unforeseen by legislators when they wrote it: protecting species requires protecting habitat. Much habitat is privately owned, and preventing degradation often means halting development, logging, and other economic activities.

Congress provided some relief to land owners in 1982 through amendments to the ESA that introduced habitat conservation plans (HCPs). HCPs encourage landowners to engage in proactive conservation efforts in exchange for some relief from ESA prohibitions. The relief comes primarily through an "incidental take" permit that exempts land owners from sanction if they accidentally harm an endangered or threatened animal while conducting otherwise lawful activities (Thomas 2001). U.S. Fish and Wildlife Service regulations implementing the law do not require that HCPs be developed with public participation beyond basic notice-and-comment procedures (Anderson and Yaffee 1998). However, habitat often is spread through multiple tracts of public and private property; as a result, many interests have a stake in the plan and therefore a seat at the negotiating table.

To bring our focus to the context of public participation, we compare two HCP cases. In 1993, the Georgia Department of Natural Resources (DNR) and the Georgia Forestry Commission (GFC) established a stakeholder-based steering committee to negotiate an HCP for red-cockaded woodpeckers across the entire state of Georgia (Crismon 1998a). The plan would apply to any private landowner in the state with small, isolated populations of the woodpecker on his or her property. The woodpecker favors old-growth, fire-dependent pine forests and is threatened by forest fragmentation and fire suppression to such an extent that the bird has been called the "spotted owl of the South."

Along with the DNR and the GFC, the committee included representatives from other federal and state agencies, timber and agricultural interests, and environmental organizations. Despite high stakes on all sides, the members of the committee were able to reach compromises for the statewide plan on several contentious issues. One participant described the negotiation process as "true give and take," although another described it as "a little testy at times" (Crismon 1998a, 6). All agreed that, once completed, the process had improved relationships among government, business, and environmental interests and had opened up new lines of communication.

The agreements reached in Georgia contrast with the outcome of an HCP effort for Tulare County, in California's Central Valley. Beginning in 1991, the Tulare County Association of Governments convened a public advisory committee to assist in the development of a regional HCP (Merrick 1998c). The plan sought to protect various resident endangered species (including the San Joaquin fox, the blunt-nosed leopard lizard, and the giant kangaroo rat) on the small amount of remaining natural habitat interspersed among the county's large agricultural acreage. Members of the advisory committee described the ensuing seven years of negotiations as "hostile," "difficult," and "contentious" (Merrick 1998c, 4). Interactions worsened over time, as participants became increasingly inflexible. Finally, the advisory committee disbanded without completing the HCP.

Where should we look to find out why discussions in the Georgia case were "testy" but ultimately productive, whereas discussions grew more hostile over time and ultimately failed in Tulare? The answer does not lie in the type of process used, because the efforts involved similar types of interest group negotiations, and they sought to produce the same type of agreement.

The important differences between the cases may lie in the contexts in which participation took place. Although both cases had the same focus on habitat management planning and were operating under the same regulatory program, there were some significant differences. The Georgia plan set policies for the diverse land uses of an entire state, whereas the California plan covered a single county dominated by agriculture. The Georgia plan concerned only one species, whereas the California plan was designed to protect more than three. Biological information was regarded as sufficient for the Tulare plan, but information about "demographic isolation" was found to be lacking as the process proceeded in Georgia. Although the case studies did not provide much information on preexisting relationships among parties, they may have been much worse in one case than in the other. The list goes on and on.

In this chapter, we examine how such contextual factors affect the success of public participation. We focus on three key factors: issue type, levels of preexisting conflict and mistrust, and institutional context (i.e., differences across local, state, and national decisionmaking processes, or across agencies).

Of course, many other contextual issues may be important, and we outline important areas for future research.

Types of Issues

Certain kinds of environmental issues may be less conducive to public participation than others. Some issues may be technically complex or involve a broad range of conflicting interests. Sometimes gains for one interest group are losses for another, making win–win solutions difficult. Other times, trades can be made among several issues that leave all parties better off. We distinguish among the cases in three ways to examine how the type of issue under debate may affect the public participation process and its results.

The first distinction is whether the case concerns overarching policy-level issues or narrow site-specific issues. We define *policy issues* as state and national decisions that affect the public as a whole. (In some cases, decisions concerned local policy issues—such as setting management policy for a park—and we considered such cases site-specific.) Cases that involved policy issues make up around 16% of the data set, and they cover such topics as the development of regulations or the identification of environmental protection priorities for a state or region. The remaining 84% of cases concerned a single site or geographic feature and tended to engage a smaller set of more specific actors. Site-specific cases involved cleaning up contaminated land, siting industrial facilities, and a range of other issues.

A second way to categorize the cases is by whether the issues involve natural resources or pollution. Cases that involved natural resources (e.g., habitat conservation planning, wildlife management, or energy policy) make up 56% of the data set. Pollution cases (e.g., water quality or the development of regulations concerning hazardous chemicals) make up the remaining 44% of the data set. Differences between natural resources and pollution issues would be expected for various reasons. Their regulatory and management regimes stem from different historical traditions and are defined by different sets of laws and institutions. Natural resources debates often concern issues of private property rights and economic development, whereas pollution controversies often revolve around concerns about health and technological change.

A third way to characterize the issues is by specific issue type. Each case fell into one of six categories:

- design of regulations and standard setting (5% of cases),
- design of permits and operating requirements (10% of cases),
- natural resources planning and management (31% of cases),
- investigation and cleanup of hazardous waste sites (22% of cases),
- facility siting (15% of cases), or
- policy development (local, state, and federal) (17% of cases).

Several hypotheses suggest why some categories might be more difficult for participation than others. Cases in which human health risks already exist, such as hazardous waste investigations and cleanups, might pose more problems for public participation than cases in which the potential for risk is a future concern. Cases in which "winners" and "losers" are clear—such as facility siting cases—also might be particularly challenging.

Regardless of how we distinguish among environmental issues, little relationship with success emerges (Figure 4-1). Policy-level and site-specific cases have similar records of success. Pollution and natural resources cases have similar records of success as well. (The statistical results for these and subsequent bivariate comparisons are presented in Appendix C.)

More difference is apparent when we compare the six specific issue types. Facility siting and policy development cases received lower scores than the

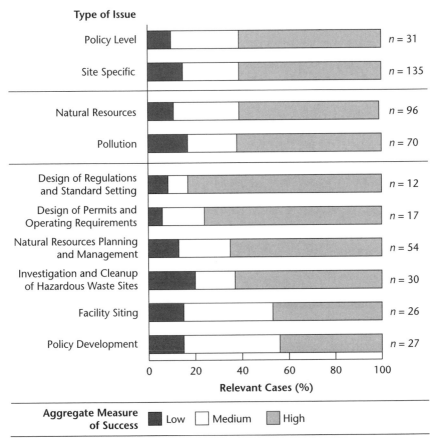

FIGURE 4-1. Aggregate Measure of Success, by Type of Issue

Note: n = total number of cases scored.

others, whereas cases of regulation and permit design scored somewhat higher. However, this result has less to do with differences among issue types than with the kinds of public participation processes used. High performers are characterized by their use of more-intensive mechanisms for involving the public (such as negotiations and mediations) (Figure 4-2). (The meaning of *intensive* and the relationship between success and the intensity of participation mechanisms are discussed extensively in Chapter 5.) When we controlled for different features of the public participation process in the multivariate analysis (Appendix C), issue type appeared to play a very limited role in whether public participation was successful.

Preexisting Relationships

A second contextual issue that may affect participation is the history of relationships among citizens and between the public and government. Not much imagination is required to understand that a process that brings together parties who have been fighting for a decade would be more challenging than one in which all parties get along. Likewise, participation can be difficult when

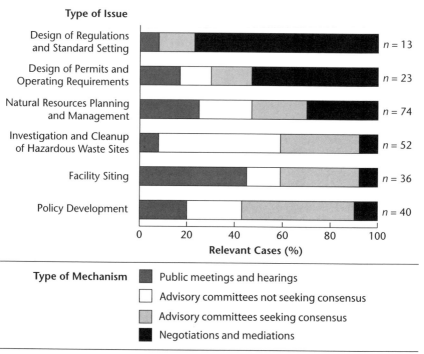

FIGURE 4-2. Relationship between Type of Mechanism and Type of Issue
Note: n = total number of cases scored.

the lead agency is deeply distrusted. Peelle et al. (1996) and Gould (1991b) argue that good preexisting relationships are conducive to public participation success. Landre and Knuth (1993) suggest that an agency's reputation is also an important factor for success.

Of the cases in which relevant data are available, 70% are characterized by a high degree of mistrust of agencies and 42% are characterized by a high level of preexisting conflict among parties. Despite the plausibility of expectations that the quality of preexisting relationships may influence results, the case study data show little correlation between success and preexisting conflict or preexisting mistrust. The multivariate analysis (Appendix C) further supports the absence of a link between preexisting relationships and success.

Evidence of a moderate relationship between good preexisting relationships and successful public participation did appear when we controlled for possible bias (detailed in Appendix D). However, this relationship was mainly a factor of the kinds of participatory mechanisms used in the unbiased and potentially biased cases rather than an artifact of actual bias. We infer that preexisting conflict and mistrust have more impact on the success of public participation when the public participation processes are less intensive. In other words, robust participation processes do a better job of transforming poor preexisting relationships than do less robust processes, but a history of conflict is not itself a significant barrier to the prospects of success.

Institutional Context

The identity of the lead agency and its level of government might play a role in explaining the success of stakeholder processes. Agency reputation, organizational culture, funding, capacity, and various other factors may be important.

State and local decisionmaking is likely to increase in importance as local issues (such as land use) come to the forefront of environmental concern. Stakeholder participation at the state and local levels is likely to increase as well. The Western Governors' Association has adopted collaboration among stakeholders as one of the core principles of its "Enlibra" doctrine, which outlines a vision for environmental policymaking in the states.

Of the cases studied, local agencies took the lead in 17% of the cases, and state governments took the lead in 38% of the cases. Combined, the state and local cases covered 40 states. Federal agencies took the lead in 38% of the cases, and a mixture of federal and state government institutions took the lead in another 7% of the cases.

No difference is apparent in the success of cases led by federal, state, or local lead agencies in the bivariate comparison (Figure 4-3). In the multivariate analysis (Appendix C), federal cases fared worse than cases in which state and local agencies led, all else being equal.

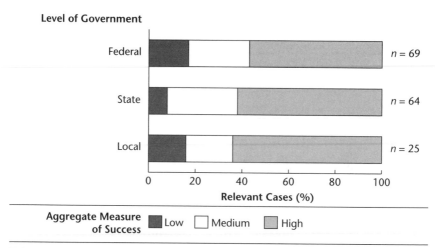

FIGURE 4-3. Aggregate Measure of Success, by Level of Government

Note: n = total number of cases scored.

For federal agencies only, the degree of success varied somewhat depending on which agency or department took the lead. Almost all federal cases fell under the jurisdiction of the U.S. Environmental Protection Agency (EPA), the Forest Service, the U.S. Department of Energy (DOE), the U.S. Army Corps of Engineers (the Corps), or resource management agencies within the U.S. Department of the Interior (DOI). As shown in Figure 4-4, EPA and DOI had more successful sets of cases than DOE, the Forest Service, and the Corps. Other than EPA, however, the number of cases for each institution is relatively small, and the results are only suggestive.

In the majority of cases (72%), the term *lead agency* means exactly what it sounds like: a government agency directly leading the process. However, in 28% of the cases, agencies had either delegated responsibility for running the process to another institution or were simply the audiences for decisions or policy recommendations generated by decision processes largely outside of government oversight and responsibility. Although these distinctions affect our conclusions about some aspects of the public participation process in Chapter 5, they make little apparent difference in the success of the cases.

Context and Success

Differences among environmental issues, preexisting relationships, and institutional contexts all appear to play surprisingly small roles in determining whether public participation is successful. These contextual issues certainly

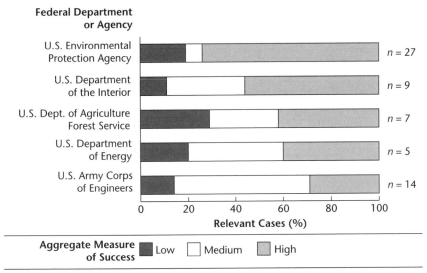

FIGURE 4-4. Aggregate Measure of Success, by Federal Department or Agency

Note: n = total number of cases scored.

play a role in how participatory processes play out, but they do not appear to predetermine outcomes.

Much more work on the role of context can be done, and much of it can use data from our case study database. A short list of priorities for future research includes examining the complexity of decisions, the degree of scientific uncertainty, the geographic scale of the issue, the complexity of jurisdictional authority, and the degree of programmatic authority.

Chapter 5

The Process of Public Participation

M any conflicts over large-scale infrastructure projects took place in the late 1960s and 1970s as citizens became more concerned with environmental damage and demanded greater input into government decisionmaking. The 1969 National Environmental Policy Act mandated environmental impact assessments for large government-led projects and gave citizens broad administrative and legal powers to challenge and influence government decisionmaking. Increasingly, citizens confronted—and confounded—government bureaucracies whose traditions were based on the construction of power plants, dams, highways, and other large-scale projects.

One of these bureaucracies was the U.S. Army Corps of Engineers (commonly known as the Corps). In its 200-year history, the Corps has been responsible for many civil works. One of its primary missions since the early 1900s has been navigation and flood protection through large-scale water projects.

We introduce our discussion of the process of public participation by contrasting different approaches taken by the Corps in two water project cases (Mazmanian and Nienaber 1979). In the first case, the Corps's resistance to anything more than minimal levels of participation accelerated conflict. In the second case, an open planning process led to widespread satisfaction with the outcome and a boost in the Corps's reputation.

The first case concerns the Corps's efforts in the 1960s and early 1970s to build a lock and ship canal in the Mississippi River Delta, near New Orleans, Louisiana. The project was a key element of a major new shipping facility and was considered by proponents to be a crucial step in maintaining the region's economic competitiveness.

However, the proposed canal and lock system would cut through St. Bernard Parish, Louisiana. Fearing the partitioning of their community, increased exposure to flooding, and damage to the local ecology, the citizens of St. Bernard Parish rose up to oppose the plan. The controversy that followed pitted local residents, local political and administrative leaders, and environmentalists against the Corps, which was backed by politically powerful economic interests, the mayor of New Orleans, and the area's representative to the U.S. Congress.

The Corps adopted its traditional "decide-announce-defend" approach to public participation. With all the major decisions already made, it convened a public meeting to inform the community about its plans for the canal. The well-attended public meeting devolved into an acrimonious spectacle. To the citizens, it was a charade—they understood that the Corps had already made its decision to go ahead with the project and that it was largely doing the bidding of powerful local economic interests. The public meeting only served to demonstrate the intensity of the conflict and the seeming impossibility of finding a solution that could accommodate the conflicting interests. When an elected official spoke in support of the Corps's project, he was shouted down with "catcalls and boos, followed by a parade of demonstrators marching around the meeting room" (Mazmanian and Nienaber 1979, 92).

Meanwhile, the Corps had another project under way near San Francisco, California. The project involved developing and implementing a plan for flood control for a low-income community that was located in the floodplain of two creeks and was subject to frequent inundations. Controlling the damage from periodic floods was considered a key prerequisite for the area's economic revitalization.

Rather than proposing a final flood control plan and bracing for community reaction, the Corps engaged in a participatory planning process. Corps staff met with organized community interests as well as local and state officials. The meetings were designed to develop a menu of options for flood control and then settle on one solution agreeable to all stakeholders. The Corps played a large role in identifying different options but also started the participatory process early enough that citizens could suggest new alternatives.

On the basis of the information generated from the informal meetings, the Corps developed a plan that responded to the desires of the various community interests. When it held an official public meeting to announce the project, conflict over various options had already been worked out. The plan "offer[ed] something to everyone" and met with universal approval (Mazmanian and Nienaber 1979, 111). It was adopted in 1975 by Corps headquarters, approved by the administration, and funded by Congress in 1976. The Corps not only had developed an acceptable and feasible plan but also had been able to establish a rapport with the community, earning trust and respect despite some initial skepticism.

To some extent, the context of the California case was more conducive to success than that in Louisiana. In California, the community would benefit greatly from a flood control project, whereas in Louisiana, the canal was perceived to convey little local benefit. Also, in California, no dominant local economic interest was pushing the project and there was less of a local history of conflict with the Corps.

What really differentiates the cases, however, are the processes used for public participation. In the California case, the Corps sought public involvement in the early stages of project planning rather than seeking public approval of a plan already devised. It sought input from all segments of the community rather than making back-room deals with economic or political interests. Most importantly, it adjusted the project plan to serve the needs of the community rather than forcing on them a project devised by bureaucrats.

In this chapter, we examine the relationship between the public participation process and success. We look at process in two ways. First, we analyze how different kinds of public participation processes relate to success. We identify four categories, which we refer to as *mechanisms,* that range along a scale of intensity from public meetings and hearings at one end to formal negotiations and mediations at the other. In both of the Corps cases, for example, the mechanism was public meetings and hearings. Second, we examine how success relates to four characteristics that vary even for a particular mechanism type: the responsiveness of the lead agency, the motivation of the participants, the quality of deliberation, and the degree of public control. It is in these attributes—what we call *variable process features*—that the two Corps cases differed significantly.

Participatory Mechanisms

Public participation processes can be organized into four categories of mechanisms. Roughly one-quarter of our cases fell into each category:

- public meetings and hearings (21%),
- advisory committees not seeking consensus (25%),
- advisory committees seeking consensus (30%), and
- negotiations and mediations (24%).

The mechanism types are not rigid definitions of rules and procedures. Rather, they are constellations of features that together define each category. They differ according to how participants are selected, who participates, how decisions are made, and what kind of output they produce (Table 5-1). The mechanisms differ in other consistent ways as well, such as the skills and training that typical participants bring to the process and the resources that the process requires.

TABLE 5-1. Participatory Mechanisms

	Feature			
Mechanism	Selection of participants	Type of participant	Type of output	Seek consensus?
Public meetings and hearings	Usually open access; group size ranges widely	Average citizens	Information sharing	No
Advisory committees not seeking consensus	Small group of participants selected based on characteristics	Average citizens, interest group representatives	Recommendations to agency	No
Advisory committees seeking consensus	Small group of participants selected based on characteristics	Average citizens, interest group representatives	Recommendations to agency	Yes
Negotiations and mediations	Small group of participants selected based on characteristics or specific interests	Interest group representatives	Agreements among parties	Yes

Mechanism Types

Public Meetings and Hearings. Twenty-one percent of the cases involve public meetings and hearings as well as related processes, such as public workshops. In most of these cases, access is open; any interested citizen can participate. Whereas participants may identify with major interest orientations—such as pro-environment, pro-business, anti-tax—or be members of interest groups, their roles are those of individual citizens, not formal representatives of groups. These processes mainly involve information exchange: agencies inform citizens about their activities, and citizens provide input and individual opinions about agency policy. Agencies are under an implicit obligation to review information from these processes, but in most cases, the commitment to sharing decisionmaking authority with the public is weak.

Advisory Committees. Advisory committees are categorized as "not seeking consensus" or "seeking consensus." In 25% of the cases, participants are not making decisions by consensus; in 30% of the cases, participants seek consensus.

Beyond how participants make decisions, the two kinds of advisory committees share many similarities. They typically have a defined and consistent membership. In most cases, project planners select participants to represent various interest groups or points of view. Sometimes they select participants to be "representative," that is, a microcosm of the socioeconomic characteristics and issue orientations of the public in a particular area. The work of advisory committees typically takes place in ongoing, regular meetings, often over years. Usually, the outcome of advisory committee work is a set of recommendations to a lead agency. Advisory committee mechanisms include citizen juries, policy dialogues, and other processes that, while not explicitly "advisory committees," share many common features with them.

Whether advisory committees seek consensus is an important distinction. In contrast to public meetings, a large part of the work of advisory committees involves managing interactions among participants (who frequently bring very different views to the table) as well as providing input to a lead agency. The procedures that guide that interaction are therefore important. Consensus requires opposing interests to work together to come to a common and acceptable solution in ways that voting and other approaches to decisionmaking do not. Consensus-based decisionmaking takes on aspects of internal negotiations among participants, and these kinds of cases commonly are facilitated by a third party. Recent analyses of public participation have challenged the wisdom of emphasizing consensus over other approaches to reaching a decision, and the attention to consensus here can help inform that debate (Gregory 2000; Coglianese 1999b).

Negotiations and Mediations. The final 24% of cases involve negotiations and mediations, in which participants form agreements that bind their organizations to particular courses of action. In some cases, the negotiating parties will implement the agreement themselves, as with many watershed management groups. In other cases, parties agree to bind themselves to a decision in exchange for a strong commitment that a lead agency will act on it. The participants in a negotiation or mediation are typically professional representatives of organized interest groups or other entities. They speak for the views of those they represent and make commitments on their behalf. By definition, these processes require decisions to be made by consensus.

Intensity of Participation Processes

The mechanisms become more "intensive" as they progress from public meetings and hearings to advisory committees to negotiations and mediations. As mechanisms become more intensive, they become less oriented toward gathering information from a wide range of people and more oriented toward forging agreements among a small group of defined interests.

As participatory mechanisms increase in intensity, two other features change as well. Data from the case studies show that participants in more-intensive processes have a greater degree of what might be called *capacity* than participants in less-intensive processes. They have more experience with the issues under discussion, more experience influencing public decisionmaking, and more experience with participatory efforts. All these skills may make these participants more effective in participating, solving problems, and getting decisions implemented. More-intensive mechanisms also require greater funding and staff support than do less-intensive processes. For example, whereas only 18% of public meetings and 18% of advisory committees not seeking consensus use a facilitator, 34% of advisory committees seeking consensus and 70% of negotiations and mediations do so.

Relating Success and Mechanism Type

Across all the cases, success correlates strongly with the intensity of the mechanism type (Figure 5-1). As the kind of mechanism used becomes more intensive, more and more cases are successful. This trend is strong and statistically significant.

One observation from Figure 5-1 deserves particular note. Little difference is apparent between the success of the two kinds of advisory committees. The use of consensus distinguishes one category from the other, and results suggest that consensus in and of itself may not play much of a role in determining which processes will be successful. Although the role that consensus plays merits additional investigation, particularly as it relates to implementation, the data provide some support for recent challenges to the wisdom of seeking consensus as an explicit goal of public participation.

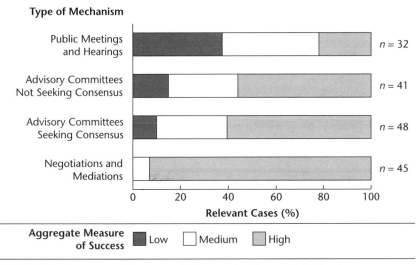

FIGURE 5-1. Aggregate Measure of Success, by Type of Mechanism

Note: n = total number of cases scored.

Given the strong relationship between success and mechanism intensity, the temptation is to conclude that negotiations or mediations should be the preferred method for involving the public in all environmental decisions. Beyond the obvious issues of cost, however, several issues suggest that this temptation should be resisted.

Although more-intensive processes appear to be better able to achieve the five social goals than less-intensive processes are, they are far less likely to meet the goals in a way that engages the wider public. Participants in more-intensive processes are less likely to reflect the socioeconomic characteristics of the wider public than those in less-intensive processes. In 79% of the cases that fell into the two more-intensive categories, participants were not socio-economically representative of the wider public. In the two less-intensive categories, this percentage was only 37%.

Perhaps more significantly, more-intensive processes are less likely to get input from the wider public through surveys, consultations, or supplementary public meetings than less-intensive processes are. Participants and agencies obtained broad input in only 22% of the consensus-based advisory committees, negotiations, and mediations, whereas 55% of the less-intensive participatory mechanisms used wider consultation. Educational outreach to the wider public also was more limited in more-intensive cases than in cases of less-intensive participation. Only 15% of the negotiation and mediation cases with adequate data demonstrated effective outreach.

More-intensive public participation processes also demonstrate a strong tendency to reach consensus by leaving out participants or ignoring issues. The exclusion of certain groups, the departure of dissenting parties, or the avoidance of issues ultimately made consensus possible—or at least easier—in 33% of the 73 cases of consensus-based advisory committees, negotiations, and mediations in which conflict appeared to have been resolved. (These kinds of cases largely drive a similar finding described in Chapter 3 for the case study pool as a whole.) In short, as processes intensify, the range and representativeness of voices heard—as well as the social benefits of education, conflict resolution, and trust formation—tend to narrow down to the relatively small group of active participants.

The U.S. Environmental Protection Agency's (EPA) reformulated gasoline regulatory negotiation illustrates how intensive public participation processes can be highly effective but ultimately undermined by their failure to engage the wider public or distribute the benefits of participation beyond a small group. The aim of the regulatory negotiation was to bring business, environmental, and public interest groups together with regulators to develop rules on the use of reformulated gasoline as a means to reduce urban smog pursuant to the 1990 Clean Air Act Amendments. Even though many of the parties were traditional adversaries, over the course of the negotiations they found creative solutions together. Ultimately, they agreed on rules that were arguably more cost-effective and more satisfying to the range of interests involved than would have been likely otherwise (Weber 1998). The process "was widely hailed as a 'milestone in reconciling the automobile with environmental quality through cleaner fuels'" (Pritzker and Dalton 1995, 391).

However, the regulations ultimately generated a massive public outcry from the people who would have to pay for those regulations but were never consulted: the driving public. "An EPA hearing [in Milwaukee] resulted in the expression of tremendous anger by the consuming public against the regulation imposed by the national government. The general perception was that consumers were the victims of a rigged inside game that excluded the 'small guy,' leaving the American motoring public to foot the bill for urban smog" (Weber and Khademian 1997, 407).

As the reformulated gasoline case shows, the choice of mechanism type is not only about the effectiveness of participation or the competence of the participants. It is also about how broadly public participation should extend the benefits of participation and the channels for public input. Whether the scales should tip toward intensive problem solving by a small group or more general analysis by a broader swath of the public largely depends on the goals of the public participation effort and the nature of the problem being addressed (topics we discuss in Chapter 7).

Our challenge to project planners is to find ways to harness the problem-solving capabilities of intensive participatory mechanisms with the broad involvement of the public more often found in the less-intensive mechanisms. One approach would be a strong emphasis on the responsibility of participants in intensive processes to communicate with, and be accountable to, their broader constituencies. A second approach would be the creative combination of participatory mechanisms. A third approach may be offered by large-group deliberative processes conducted on the Internet, which are currently under experimentation (Schneider and Beckingham 2000; Beierle 2001).

Variable Process Features

Some aspects of the public participation process are relatively independent of the kind of mechanism used. They can vary from advisory committee to advisory committee, for example, or from negotiation to negotiation. In the research literature and in our previous research, we identified four important features to consider in understanding what leads a public participation process to success:

- responsiveness of the lead agency,
- motivation of the participants,
- quality of deliberation, and
- degree of public control.

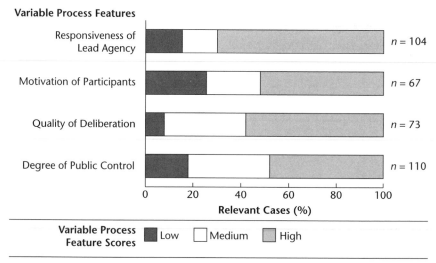

FIGURE 5-2. Variable Process Feature Scores

Note: n = total number of cases scored.

To accurately compare how these four process features relate to success, we excluded certain kinds of cases from our analysis. Because of the focus on the quality of deliberation, we included only cases that involved deliberation, and therefore we excluded those that used public meetings and hearings. Because of the focus on the responsiveness of the lead agencies, we included only cases in which a lead agency was actively involved and excluded cases that evolved outside of an agency's direct oversight. Figure 5-2 shows the scores for the four variable process features.

Responsiveness of the Lead Agency

This feature measures the interrelated aspects of agency commitment to and communication with participants. Across cases, responsiveness ranges from situations in which communication was poor and financial support was inadequate to cases in which agency personnel and participants were involved in active deliberation, resources were adequate, and high-level decisionmakers were involved. Communication between government and the public is important (Fiorino 1990), and expressions of commitment help legitimize participation in the eyes of participants (Gurtner-Zimmermann 1996). In the pilot study, commitment and effective communication were highly related to successful processes (Beierle and Konisky 2000).

Of 104 cases with relevant data, 70% received high scores for responsiveness, 14% received medium scores, and 15% received low scores. The correla-

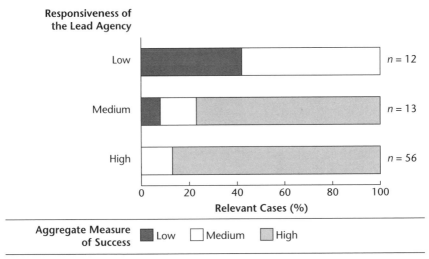

FIGURE 5-3. Aggregate Measure of Success, by Responsiveness of the Lead Agency

Note: n = total number of cases scored for both responsiveness of lead agency and aggregate measure of success.

tion between responsiveness of the lead agency and success is high, positive, and statistically significant (Figure 5-3). Low levels of responsiveness appear to foster perceptions of process illegitimacy and to lower trust.

Motivation of the Participants

This feature measures the optimism and ambition that carry the public forward in a public participation process. Across cases, differences in motivation ranged from pessimism and little confidence in the process to strong enthusiasm and a dedication to making the process work. The particular skills and abilities that people bring to the public participation process are important—including those that help them cope with adversity and keep them motivated to see a process through to completion (Hartley 1999). Other aspects of motivation arise from the process itself, such as the perceived ability to influence outcomes. In the pilot study, the ambition of participants emerged as an important factor in successful processes (Beierle and Konisky 2000).

Of 67 cases with relevant data, 52% received high scores for motivation, 22% received medium scores, and 25% received low scores. The correlation between motivation of participants and success is moderate to high, positive, and statistically significant (Figure 5-4). In cases that received high scores, participants were enthusiastic and worked hard to make the process successful. In

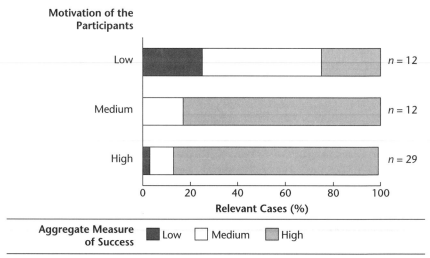

FIGURE 5-4. Aggregate Measure of Success, by Motivation of the Participants

Note: n = total number of cases scored for both motivation of the participants and aggregate measure of success.

cases that received low scores, attrition and poor attendance at meetings were common, and many groups accomplished less than expected.

Quality of Deliberation

This feature measures the quality of communication and dialogue among participants. Criteria for good deliberation include the primacy of good arguments rather than overt power, the ability to question claims and assumptions, participant sincerity and honesty, and comprehension (Renn, Webler, and Wiedemann 1995; Innes 1998). The quality of dialogue in the cases ranged from examples of persistent problems of discussion and mutual understanding to examples in which all participants engaged in easy and sympathetic discussions. In the pilot study, the quality of deliberation was highly related to the success of the process (Beierle and Konisky 2000), a finding supported by other analysts (Susskind and Cruikshank 1987; Dryzek 1997).

Of 73 cases with relevant data, 58% received high scores for the quality of deliberation, 34% received medium scores, and 8% received low scores. Charting a solid relationship between deliberation and success is difficult because our data set contained few cases in which deliberation was poor. Instead, the case sample mainly allowed a comparison between high-quality deliberation (i.e., participants have the opportunity to be heard and to understand each other) and moderate-quality deliberation (i.e., some problems or barriers to

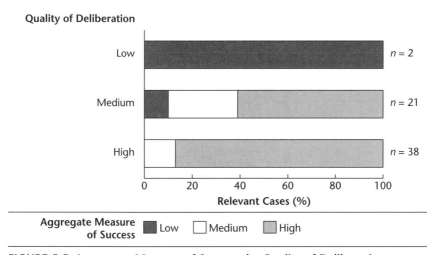

FIGURE 5-5. Aggregate Measure of Success, by Quality of Deliberation

Note: n = total number of cases scored for both quality of deliberation and aggregate measure of success.

communication persist). The relationship between the quality of deliberation and success is, however, moderate, positive, and statistically significant (Figure 5-5).

Degree of Public Control

This feature measures the extent to which participants (rather than a government agency) control the initiation, design, and execution of the public participation process. In our data set, public control ranged from cases in which citizens' influence over the process was severely restricted to cases in which the public was essentially in charge. A well-developed line of thinking in the public participation literature argues that citizen freedom and power are required for successful public participation (Arnstein 1969; Fiorino 1990; Renn, Webler, and Wiedemann 1995). Public control also has been linked to the social goal of building trust (Slovic 1993; Schneider, Teske, and Marschall 1997). In the pilot study, a high degree of public control did not necessarily relate to success (Beierle and Konisky 2000). Rather, in several successful cases, participants had relatively little freedom to design and control the public participation process.

Of 110 cases with relevant data, 48% received high scores for public control, 34% received medium scores, and 18% received low scores. The relationship between success and the degree of public control is low but positive and

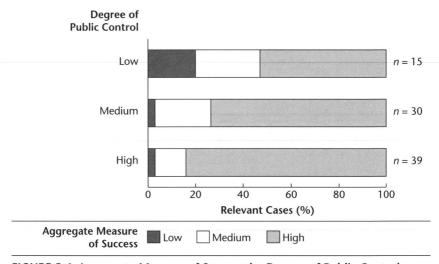

FIGURE 5-6. Aggregate Measure of Success, by Degree of Public Control

Note: n = total number of cases scored for both degree of public control and aggregate measure of success.

statistically significant (Figure 5-6). Quantitatively, this relationship is the weakest of the four variable process features.

The Importance of Process

Our analysis indicates that process has a very strong relationship to success, in terms of both mechanism type and variable process features. More-intensive participatory mechanisms are more successful across cases, although not without important caveats. Each of the variable process features correlates well with success, to a different degree. Responsiveness of the lead agency and motivation of participants top the list, followed by quality of deliberation and then degree of public control.

Variable process features tell us quite a bit more about the success of public participation than do mechanism types alone. When we control for mechanism type in the multivariate model (Appendix C), a composite score that represents the four variable process features is both positive and significant, indicating that high scores on the variable process features lead to more successful public participation processes. Adding this composite score also roughly doubles the explanatory power of the model, which means that the model does a better job of explaining why public participation sometimes succeeds and sometimes fails. Whereas the choice of mechanism type can tell us a lot about whether a process will be successful, the variable process features tell us even more.

Chapter 6

Public Participation and Implementation

One of the most pervasive arguments in support of public involvement is that, although the process may require more time initially, it leads to more effective and timely implementation later. Common sense suggests that high quality decisions that reflect public values and are made through processes that reduce conflict and mistrust should be more easily put into action.

However, some of the cases discussed thus far and the broader research literature suggest caution in making easy connections between good public participation and successful implementation. Of the two remedial action plan (RAP) cases discussed at the beginning of Chapter 3, the acrimonious Detroit River case—not the harmonious Buffalo River one—demonstrated more progress toward full implementation five to seven years after the planning phase ended. Indeed, good public participation and successful implementation have not been consistently linked in RAP cases as a whole (Beierle and Konisky 2001). One analyst of RAP cases concludes that whatever implementation has occurred has had very little to do with the participatory planning process (Gurtner-Zimmermann 1996).

Some other cases we have described tell similarly complex stories of implementation. In the case of Georgia's habitat conservation plan for the redcockaded woodpecker, participants successfully negotiated many aspects of the plan, but a national environmental group put implementation on hold by raising questions about "demographic isolation" that required additional biological studies. The U.S. Army Corps of Engineers never implemented its canal project in Louisiana, but not because of community opposition; it approved the project even after the contentious public hearing. President Jimmy Carter canceled the project years later as part of a review of federal water resource projects.

The tenuous connection between public participation and implementation in these cases is largely supported by the research literature. An exhaustive study of environmental mediation literature indicated that agreements were implemented in 80% of site-specific cases but only 41% of policy-level cases (Bingham 1986). In research on regulatory negotiations, government and interest groups were facing off in court even after they had reached consensus agreements (Coglianese 1997).

The road from policymaking to policy implementation is long and complicated. A host of political, social, and legal influences come into play. In examining implementation, we see participation in its larger context, that is, how it fits into the complex web of public policy. From this perspective, public participation appears as only part of the machinery—perhaps only a small part—that turns ideas into action.

What Do We Mean by "Implementation"?

Progress from public participation to implementation proceeds in five stages, starting with the output of the public participation process and ending with real changes in the environment (Figure 6-1).

Stage 1 is the output of the public participation process (according to the conceptual model described in Chapter 2). In most cases, the output is some sort of completed document, such as a report of findings and recommendations or an agreement among parties. In cases that use less-intensive mechanisms, output may be as simple as comments collected from the public.

Stage 2 is a decision or commitment on the part of a lead agency or other authority regarding the substance of the issue under discussion. One of the goals discussed in Chapter 3, incorporating public values into decisions, captures this stage of implementation; case studies that received high scores for this goal were those in which participants influenced decisions.

Stage 3 is a concrete change in law, regulation, or policy that officially incorporates the output of participation into legal doctrine or bureaucratic operations. Such a change might be the approval of a site for a proposed facility, the official adoption of a management plan, or the award of an operating permit. In this step, law, regulation, or policy grants official status to public input.

In Stage 4, institutions take actions on the ground: contaminated sites are remediated, facilities are built, or habitat is restored. Some implementation

Stage 1:	Output of the public participation process, such as recommendations or an agreement
Stage 2:	Decision or commitment on the part of the lead agency
Stage 3:	Changes in law, regulation, or policy
Stage 4:	Actions taken on the ground
Stage 5:	Changes in environmental quality

FIGURE 6-1. Five Stages of Implementation

actions are quite monumental, whereas others (such as additional research) are less so. Where possible, our coding accounted for the relative importance of actions taken compared with the overall task of implementation, but such information was often difficult to discern from the case studies.

At Stage 5, the final stage in implementation, environmental quality has improved—for example, rivers and groundwater are cleaner, or the population of an endangered species has recovered.

Linking Participation and Implementation

In coding a measure of implementation, we identified the highest stage on which there was information in the case study—in most cases, it was change in law, regulation, or policy (Stage 3) or actions taken (Stage 4). Accordingly, the discussion of implementation focuses on these two stages. Very few cases reported on changes in environmental outcomes (Stage 5); in most cases, such impacts are simply too far in the future to determine at the time of the analysis.

We scored the likelihood of implementation for each case. In the cases that received high implementation scores, implementation was completed or very likely; in cases that received medium scores, implementation was moderately likely; and in cases that received low scores, implementation was stalled or unlikely.

For the 61 cases in which case studies had information only on implementation as it pertained to changes in law, regulation, or policy, 70% received high scores, 15% received medium scores, and 15% received low scores (Figure 6-2). For the 90 cases in which case studies had information on implementation as it pertained to actions taken, 49% received high scores, 30% received medium scores, and 21% received low scores. Overall, the record of implementation looks rather good, but it worsens as the stage of implementa-

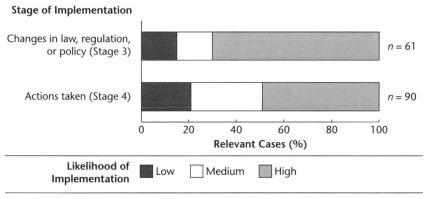

FIGURE 6-2. Likelihood of Implementation for Stages 3 and 4

Note: n = total number of cases scored.

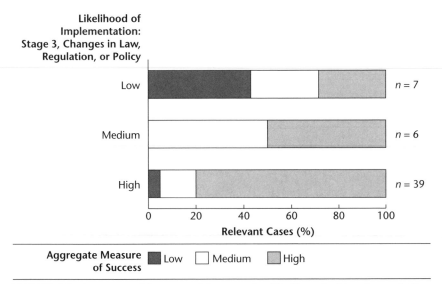

FIGURE 6-3. Aggregate Measure of Success, by Likelihood of Implementation (Stage 3)

Note: n = total number of cases scored for both likelihood of implementation and aggregate measure of success.

tion moves from the realm of law, regulation, and policy to actions taken on the ground.

The way we have treated implementation in this study is admittedly crude. It transforms the implementation process into a question of whether a definable set of stages were completed. Public participation's relationship to implementation may be much more subtle, dealing more with how options were considered, adopted or rejected, and ultimately acted upon.

Bearing that caveat in mind, however, initial insights indicate that the relationship between the quality of participation and the degree of implementation is, at most, moderate (although positive and statistically significant) for Stages 3 and 4 (Figures 6-3 and 6-4, respectively). The data on implementation and its link to successful participation are probably overly optimistic. Most case studies are written when implementation is in the early stages, if at all. Authors often refer to the likelihood of implementation, rather than to implementation itself, and the success of a recently completed public participation process may color their expectations. Also, the significance of the implementation activities reported is difficult to discern; authors may report success on the easy problems and not mention the most difficult ones. Our analysis shows that taking potential author bias into account casts the apparent link between good participation and good implementation into serious doubt: the statistical significance of results largely disappears (Appendix D).

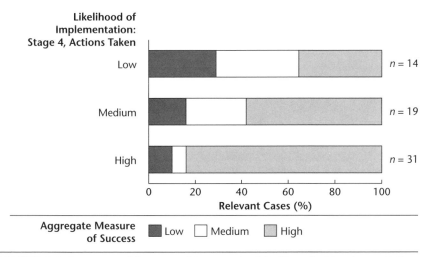

FIGURE 6-4. Aggregate Measure of Success, by Likelihood of Implementation (Stage 4)

Note: n = total number of cases scored for both likelihood of implementation and aggregate measure of success.

The analysis corroborates our intuitive sense from reading the cases that many forces unrelated to public participation influence implementation. As the stages of implementation progress toward action, a host of political, social, and technical influences come into play. Legislative changes trigger politics as usual. Regulatory changes require broader public review and comment and introduce the discipline of judicial review. Policy changes must consider the mandate of public agencies and the political climate in which they operate. Actions taken on the ground require staff and, usually, large budgets.

Forces Acting on Implementation

Information in the case studies highlighted several forces that influence implementation. Such forces may help implementation along, halt it, or make it irrelevant. Some forces are related to participation, and some are not.

Disagreements Stall Implementation

In most of the cases in the data set, participants and the lead agency agreed about what implementation steps they should take. But in some cases, they could not agree, and the debate shifted to more traditional political and legal realms. As a result, implementation was delayed or stalled entirely.

Among the 90 cases that were coded for implementation actions taken (Stage 4), the participation process ended in disagreement over what imple-

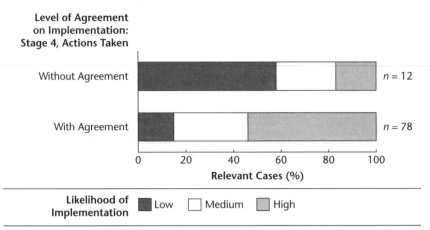

FIGURE 6-5. Likelihood of Implementation with and without Agreement between Participants and Agency for Stage 4

Note: n = total number of cases scored.

mentation should be in 12 cases. The implementation record in these cases is decidedly poor (Figure 6-5). Rather than moving on to implementation, disagreements moved into the courts, persisted as unresolved issues among regulatory agencies, led project proponents to abandon their efforts, or smothered projects under the weight of opposition. In only a few cases did a lead agency or project proponent go ahead with a project despite continued disagreement.

Failed implementation can be good news for some parties. For example, in Kern County, California, a citizens advisory committee recommended against a proposal for the expansion of a toxic waste dump. When the county government went ahead and approved the project anyway, the committee members and other citizens in the area sued the county, putting implementation of the expansion on hold (Cole 1999).

Conflict Not Really Resolved: Excluding People or Issues

In another set of cases, problems arose when issues or interests that initially had been left out of the public participation process later emerged. In several cases, conflict was reduced but not resolved by leaving controversial issues off the table or by failing to include certain parties (see Chapter 3). Such choices often come back to haunt projects in the implementation phase, as in the following two examples.

In the Three Rivers watershed of northern Ohio, the U.S. Army Corps of Engineers' plan for land disposal of sludge garnered acceptance from citizens whose wastewater treatment system would produce it (Mazmanian and Nienaber 1979). However, residents who lived near the disposal area were involved only minimally in the public participation process. These residents opposed the plan, and ultimately, the land disposal effort was shelved.

Leaving issues off the table can lead to similar results. In the development of a plan for cleaning up near-shore areas around Rochester, New York, large waste discharges from Eastman Kodak—a powerful local industry—were downplayed to ensure a more harmonious process (Kellogg 1993b). Failure to address the relationship between discharges from Kodak and the community's desire to eliminate restrictions on fish consumption was expected to become a large stumbling block for implementation.

Political Intervention

With some frequency, implementation in controversial cases is pushed forward or halted by a politician, who brings political power to bear on decisions normally handled administratively.

As mentioned earlier, President Jimmy Carter ultimately settled the dispute over the U.S. Army Corps of Engineers' canal project in Louisiana (Mazmanian and Nienaber 1979). In a case involving the damming of the Snoqualmie River in Washington, the governor repeatedly stepped in, originally opposing the dam project and then endorsing a mediated agreement forged among stakeholders (Cormick and Patton 1980). In a controversy over allowing oil drilling in Michigan's Sand Lakes Quiet Area, the support of the governor's wife helped a citizen coalition publicize the need to protect the area (Nelson 1990b).

In each of these cases, public participation may have raised the profile of an issue and brought it to the attention of powerful individuals. However, political intervention—not participation per se—determined what would be implemented.

Changing Circumstances Make Implementation Undesirable

Public participation can be a long process, and the path to implementation can be even longer. In some cases, the passage of time changes the context of decisionmaking so much that implementation of decisions forged through a public participation process no longer makes sense. Changed economic conditions, political parties, or other issues can make decisions irrelevant, inappropriate, or unsupported in the new situation.

For example, in the Western Massachusetts Electric Company Collaborative, participants sought to develop energy conservation policies. However, a rise in energy prices and looming deregulation threatened the agreed-upon policies (English and others 1994). Economics also played a part in the famous case of the Asarco smelter in Tacoma, Washington. Even though U.S.

Environmental Protection Agency (EPA) Administrator William Ruckelshaus consulted the citizens of Tacoma about whether the facility should shut down, economic issues—not the efforts of the community or EPA—eventually determined the smelter's fate (Scott 1988; Krimsky and Plough 1988).

Links to Policies and Programs

In some cases, what fosters or hinders implementation is not participation per se but the larger regulatory program in which participation and implementation operate. Program budgets, regulatory power, and staff are usually the principal drivers behind implementation, and public participation is simply one piece of a decisionmaking process along the way.

Many of the cases regarding Superfund cleanup illustrate this effect. One is the highly successful Clear Creek Watershed Forum, a Colorado group seeking to improve water quality and hazardous waste remediation in the Clear Creek watershed. The group has been described as an "effective catalyst for promoting field-level actions," including a tailings capping project, an emergency response system, and various other efforts (Kenney 1997, 27). However, agencies probably would not have undertaken any such activities had Clear Creek not been in the Superfund program. The regulatory and budgetary muscle behind the cleanup law, not the watershed forum, drove implementation.

The role that programmatic backing plays in implementing decisions is most obvious in cases where it is lacking. One extreme example is a consensus-based process initiated by DuPont to resolve controversy about the company's plans to mine titanium along the eastern border of the Okeefenokee National Wildlife Refuge in Florida (RESOLVE 1999). The process (which involved DuPont, local officials, environmental and community groups, three Native American tribes, local landowners, and mining interests) was highly successful in resolving conflict and developing an agreement among participants. However, a crucial part of the agreement called for a $90 million buyout of DuPont's land and mineral rights. Much of the bailout money would have to come from federal sources, but the federal government was not a party to the negotiating process and had not agreed to provide the money. Indeed, then-Interior Secretary Bruce Babbitt expressed "strong reservations" about the bailout agreement and called the $90 million "grossly inflated," casting implementation into serious doubt (*Georgia Times-Union* 1999, A-1).

In the Okeefenokee case and others, public participation—even when done well—is not a substitute for the regulatory power, political will, and money required to get things done.

Chapter 7

Designing Public Participation Processes

The goal of evaluation is to provide guidance for the future by understanding the past. In this chapter, we outline a method for designing future public participation efforts, applying lessons discussed throughout this book. Because public participation is an art as well as a science, we draw on our informal insights as well as our formal results. We intend for this chapter to be helpful to project planners, both inside and outside of government, who are responsible for designing an effective public participation effort.

Because each situation in which public participation is used is unique, the public participation process cannot be created from a standard blueprint. However, it is possible to apply a methodical approach to process design. All too often, participation is pursued because it is considered a good thing in and of itself, and particular mechanisms are used because project planners are most familiar with them.

We suggest a different approach. In this chapter, we describe five steps in which project planners determine why public participation is necessary, identify the goals of the process, answer design questions, select and modify a process, and follow up with evaluation.

Step 1: Determine the Need for Public Participation

In many of the cases in which it was unsuccessful, public participation probably should not have been a part of decisionmaking in the first place. Or, put another way, the agency initiating the public participation process was not willing to make the kinds of commitments that were necessary to make the process successful. The agency probably should have forgone participation altogether or, at most, complied with only the most basic statutory requirements.

Three kinds of rationales for public participation are useful for determining whether it is warranted: instrumental, substantive, and normative (Fiorino 1990; Perhac 1996). Decisionmakers should ask whether any of the rationales justify participation in their particular situation.

Instrumental rationales argue that public participation facilitates policy formation and implementation. Only by resolving conflict, building trust, or developing "buy-in" through participation can progress be made on an issue. In the case studies examined for this report, instrumental rationales were particularly critical to cases of collaborative resource management, such as the habitat conservation plan negotiations (see Chapter 4). In such cases, parties and institutions from local landowners to state agencies are needed to buy in to an agreement and assist with implementation, often voluntarily, to improve watersheds and other resources.

Substantive rationales argue that public participation leads to objectively superior decisions. The public may bring valuable information, a deeper understanding, or creative thinking to bear in solving a particular problem. In the Buffalo River remedial action plan case (see Chapter 3), regulatory personnel came to realize that the public had a wealth of local knowledge and technical capacity that regulators could use to understand water quality problems in the contaminated river (Kellogg 1993b).

Normative rationales argue that public participation is both a right of citizens and a route to a more healthy democratic society. Such rationales are most relevant when traditional decision tools are unlikely to capture the range of public values in play. Many of the case studies in which decisionmakers sought decision criteria from the public (e.g., for siting an industrial facility) explicitly acknowledged that public values had something vital to add to decisionmaking.

Beyond identifying what might provide a rationale for public participation, decisionmakers must consider whether they are willing to accept two commitments that go along with seeking participation. First, decisionmakers must commit to some degree of flexibility and open-mindedness regarding the nature of the process and its outcomes. Participation shapes participants' understandings, attitudes, and expectations. Participants may want to redefine problems, focus on different issues, or otherwise change the nature of questions that agencies ask. Measures of the responsiveness of the lead agency (which was highly related to success) captured how agencies responded to these requests. Second, decisionmakers must recognize the legitimacy of public values and understand that those values may lead to priorities and conclusions that agencies (which have their own understanding of what the public interest is) find wrong. Failure to make these two commitments threatens the legitimacy of the public participation process and whatever public trust the lead agency may have.

Step 2: Identify the Goals of the Process

After it has been determined that public participation seems warranted, the goals of the lead agency and of the public must be considered. Every process has some specific goals, such as solving a particular problem or producing a set of recommendations. But some or all of the social goals discussed throughout this book usually apply as well. Will the process need to identify public values and incorporate them into decisions? Are education and problem solving important? Does the process need to resolve conflict? Does it need to build trust?

In identifying goals, possible barriers to their achievement should be considered. Such barriers mainly fall into what we described earlier as the "context" of decisionmaking. For example, if the degree of preexisting conflict or mistrust over an issue is high, then these conditions must be addressed in the process. Similarly, complex topics need processes that emphasize education.

The goals will dictate many of the design features of the public participation process and provide initial insights into the kind of mechanism that is most appropriate. Figure 7-1 shows how different mechanisms achieve different goals.

The data in Figure 7-1 support our earlier conclusion that more-intensive mechanisms tend to perform better than less-intensive mechanisms across the range of social goals. But differences in goal achievement within mechanism categories are also important. For example, public meetings and hearings were more effective in improving the substantive quality of decisions than in resolving conflict. Figure 7-1 is a rough first step toward designing a process, one that is refined in Step 3.

Step 3: Answer Design Questions

Project planners should answer four design questions to refine their decision about what kind of participatory mechanism to use: who should participate? what kind of engagement is appropriate? how much influence should the public have? and what role should government play?

Who Should Participate?

One of the most challenging questions in public participation is, who is the public? The obvious answer is, we all are. However, in any conceivable participation process (except perhaps voting), project planners call on a relatively small group of people to act as a proxy for the larger public. Two key considerations address how participants should represent the broader public.

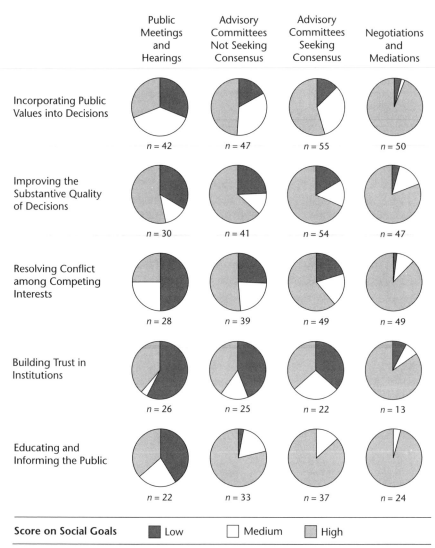

FIGURE 7-1. Success in Meeting the Five Social Goals, by Mechanism Type

Note: n = total number of cases scored.

First, determine how far the reach of participation should extend. Issues that affect a broad section of the public demand broader participation than issues that affect a relatively narrow set of private interests. For wide-ranging processes, the benefits of participation—in terms of educating the public, resolving conflict, or building trust—must be spread widely. An apparent trade-off exists among kinds of participatory mechanisms (see Chapter 5). Whereas more-intensive mechanisms may be more effective than less-intensive mechanisms at solving problems among a small group, they often exclude the wider public.

Second, determine what kind of representation is desirable when small groups represent the wider public. This consideration, like the first, also involves a trade-off. Project planners usually are faced with the choice between participants who are representative in terms of the various interests affected by an issue or representative in terms of socioeconomic characteristics (e.g., age, race, education, and income). For example, participants in regulatory negotiations are selected to represent broad categories of interest groups, whereas participants in citizen juries are randomly selected to reflect the socioeconomic characteristics of an area.

Choosing between the two kinds of representation is essentially a decision between instrumental and normative rationales. Interest group representatives are more likely than average citizens to wield the political influence that makes sure their input is heard and acted on. However, average citizens probably reflect broad public values more accurately. The specific goals of the process usually determine which alternative is most appropriate.

What Kind of Engagement Is Appropriate?

The kind of engagement chosen for a public participation process largely comes down to a choice between information sharing and deliberation. *Information sharing* is the two-way exchange of information that might take place in a public meeting or public hearing. Citizens hear what agencies are doing, and agencies hear what citizens think of their plans. *Deliberation* makes this two-way exchange more iterative and intensive, stressing well-reasoned arguments and group problem solving.

Participatory mechanisms based on information sharing have some distinct advantages: they are relatively inexpensive, easy to deploy, and usually do not require a large time commitment on the part of agencies or the public. In many cases, they are sufficient for meeting goals oriented toward informing citizens and agencies. However, if the set of goals is more complex—particularly if it involves resolving conflict or building trust—then information sharing is likely to be insufficient.

Deliberative processes are much more likely than information-sharing processes to generate successful outcomes on a wide range of social goals. They create a forum for participants to discuss their values and ideally identify areas of common ground. Democratic deliberation may lead participants to identify community-oriented values rather than individual values and create an atmosphere for problem solving that can improve the substantive quality of decisions (Dryzek 1997). Deliberation can be the key to resolving conflict because it brings competing stakeholder groups together to work out their differences face to face, rather than leaving agencies to arbitrate disputes.

Deliberation does not necessarily require a search for consensus, but many deliberative processes are consensus-based. The records of success between advisory committees that seek consensus and those that do not are not significantly different (Chapter 5). The choice of whether to seek consensus probably should be an instrumental one: can implementation move forward given only one agreed-upon solution, or can agencies and the public tolerate more than one possible solution?

How Much Influence Should the Public Have?

Whereas a public participation process implicitly requires some level of public influence, the amount of influence ranges from providing information to formulating recommendations to forging agreements. In most public meetings, the public provides only information and comments; the agencies have little obligation to act on the public's contributions. In contrast, most regulatory negotiations involve an explicit commitment by agencies to adopt the resolution of the negotiation as an official proposed rule. How should planners determine the extent of the public's influence?

The first relevant consideration is the motivation of participants. Participation demands a lot from people. For agencies that hope to capitalize on the abilities and motivation of the public, the relevant currency is influence. In the case studies, one of the principal reasons offered for low levels of participant motivation was a perception that the public had little influence over agency decisions. Planners must consider how much influence must be granted to get participants to accomplish the goals of the process.

The second issue to consider is building trust. Some research suggests that one of the few ways agencies can rebuild trust is by increasing public influence on decisionmaking (Schneider, Teske, and Marschall 1997; Slovic 1993). In our analysis, the goal of incorporating public values, which essentially measures the public's influence, is highly and significantly correlated with the goal of building trust (Appendix B). In low-trust situations, then, the public may need to be granted more influence to convince them of the legitimacy of the public participation process.

What Role Should Government Play?

Agencies can decide how involved they should be in organizing and running a public participation process. The choice relates to our examination of responsiveness of the lead agency and degree of public control (Chapter 5). Responsiveness is important to a successful process, especially for creating trust in the lead agency. However, to say that agencies should be responsive is not to say that they should control the process. The results of our analysis indicate that too much control—to an extent that agencies suppress the public's ability to affect decisionmaking—undermines successful participation.

The role of government agencies in public participation processes, then, is a balancing act between responsiveness and control. As processes become more intensive and participants take on greater responsibility for the content of decisions, the role of government generally shifts away from active leadership. Although agencies may be active as participants in more-intensive processes, they often relinquish much of the control of the process to a facilitator or to the group itself.

Step 4: Select and Modify a Process

Answers to the four design questions above will largely define the kind of participatory mechanism that is most appropriate. The design questions can be recast as five trade-offs:

- *scope of inclusion:* narrow versus broad
- *representation:* socioeconomic versus interest group
- *kind of engagement:* information sharing versus deliberation
- *level of public influence:* limited versus moderate or high
- *role for government:* passive versus active

Various participatory mechanisms are classified in Table 7-1 according to these trade-offs. In addition to the mechanisms discussed throughout this book, public comments, surveys, and citizen juries are included to illustrate a broad range of alternatives.

Project planners may find that a single mechanism suits their needs. However, a single mechanism probably will be only approximately right, because the tasks and demands of decisionmaking change over time. Fortunately, mechanism types are not cast in stone; they are simply conglomerations of design choices that can be modified to suit particular needs. Processes can be combined in ways that help meet all of the goals of the participatory effort.

We would be remiss if we did not mention a final issue that figures prominently in most planning processes: cost. Clearly, less-intensive participatory mechanisms cost less than more-intensive processes in terms of finances and staff time. Less-intensive processes also demand less of participants' limited time and energy. Cost considerations may limit planners to a process that is less intensive than ideal. If this is the case, the goals of such a process must be tempered and the expectations of agencies and participants clarified in light of limited resources. It is far preferable to reevaluate goals from the start than

TABLE 7-1. Design Features of Mechanism Types

Type of mechanism	Scope of inclusion		Representation		Kind of engagement		Level of public influence		Role of government	
	Narrow	Broad	Socio-economic	Interest group	Information sharing	Deliberative	Low	Moderate or high	Passive	Active
Public comments		X	X		X		X			X
Surveys		X	X		X		X			X
Public meetings and hearings		X	Varies	Varies	X		X			X
Advisory committees not seeking consensus	X		Varies	Varies	Varies	Varies	Varies	Varies	Varies	Varies
Advisory committees seeking consensus	X			X		X		X	Varies	Varies
Citizens juries	X		X			X	X		X	
Negotiations and mediations	X			X		X		X	X	

to pursue an ambitious set of goals using a circumscribed process that stands little chance of achieving them.

Step 5: Evaluate the Process

One of the benefits of a well-thought-out project planning process is its ability to facilitate evaluation. Analysts can turn the goals identified in Step 2 into evaluation criteria. The evaluation then can test the assumptions that drove design choices in Step 3 (e.g., whether the right people were selected to participate and whether the public had the right level of influence).

We strongly recommend that an evaluation be budgeted into every project to help project planners better understand what the public and agencies accomplish and to build an information base of the approaches to public participation that achieve satisfactory results.

Examples of Project Planning

To illustrate the planning process, we compare three types of projects: one that involves fact finding, one that involves visioning and setting goals, and one that involves implementation planning. These different projects could just as well be three phases of one project. The design steps for each of the three projects are given in the table on the next page.

The rationale and goals (Steps 1 and 2) for each project are quite different. The rationale for the fact-finding project is substantive; its goal is to gather the best information and ideas from the broadest swath of possible sources. The rationale for the visioning project is normative; its goal is to have people reflect on what they want for the place where they live. The rationale for the implementation planning project is mainly instrumental; the goals are to implement the project and to reduce the conflict and mistrust that might inhibit implementation. Different rationales and goals lead to different answers to the design questions (Step 3) and, in turn, to different suggested processes (Step 4) and different evaluation criteria (Step 5).

The fact-finding project seeks to involve as many people as possible who have relevant information. Engagement will involve mainly sharing information, and the public's influence is predicated on the quality of its contributions. The government's role is significant, defining what areas of information are needed and how they will be used. Traditional public participation processes (e.g., informal consultations, notice and comment procedures, and public meetings) probably will be sufficient.

continued on next page

Examples of Project Planning—continued

TABLE. Planning Public Participation for Three Types of Projects

Step	Project or phase		
	Fact finding	*Visioning and setting goals*	*Implementation planning*
Step 1: Determine the need for public participation	Substantive rationale	Normative rationale	Instrumental rationale
Step 2: Identify the goals of the process	Increase the substantive quality of decisions	Identify and incorporate public values into decisions	Reduce conflict, build trust, implement decisions
Step 3: Answer design questions			
Who are the participants?	Everyone	Interested citizens	Interest groups
What type of engagement is appropriate?	Information sharing	Deliberation	Deliberation
How much influence should the public have?	Input	Recommendations	Agreement
What role should government play?	High control	Moderate control	Low control
Step 4: Select and modify a process	Public comments; public meetings and hearings	Series of advisory committees not seeking consensus; workshops; citizens jury	Advisory committee seeking consensus; negotiation and mediation
Step 5: Evaluate the process	Were decisions based on better information or ideas?	Was a common vision formulated among participants and the wider public? Was the vision incorporated into decisions?	Was conflict resolved and trust increased? Were decisions implemented?

Examples of Project Planning—*continued*

The visioning project requires a different kind of process. Because broad public values are at stake, a broad and representative group of participants is required. But because the participants must discuss and debate how competing sets of individual values will arrive at a collective vision, the process probably needs to be deliberative. Deliberative processes generally require small groups, so a series of workshops or small-group discussions among all those interested enough to participate is desirable. The public's influence will be largely through recommendations made to an agency; government will play a less visible role, allowing the deliberations to evolve without overt control.

The implementation project involves a very different structure. The preferred participants probably are no longer the public at large but the groups whose interests will be directly affected by the outcome and who will play a significant role in implementation. The process needs to encourage creative problem solving and thus should emphasize deliberation and ensure that the participants have access to high-quality information and analysis. In many cases, the participants will have a great degree of influence, perhaps even forging agreements among themselves regarding each participant's implementation responsibilities. Government may play a very hands-off role, providing only an ultimate objective, an implicit assurance to go along with the participants' agreement, and technical resources.

Each kind of project dictates very different approaches to public participation rather than a one-size-fits-all approach that would be appropriate for, at best, one out of the three projects. A consideration of rationale, goals, and design leads to three distinct process designs that have a much higher likelihood of success.

Chapter 8

Conclusions and Areas for Further Research

The case study record of the past 30 years paints an encouraging picture of public participation. Involving the public not only frequently produces decisions that are responsive to public values and substantively robust, but it also helps to resolve conflict, build trust, and educate and inform the public about the environment.

In understanding what makes participation successful, process is of paramount importance. More-intensive mechanisms generally are more successful than less-intensive mechanisms. Processes in which agencies are responsive, participants are motivated, the quality of deliberation is high, and participants have at least a moderate degree of control over the process are more successful than processes that do not have these characteristics. Good processes appear to overcome some of the most challenging and conflicted contexts.

Much can be done to improve the practice of public participation. Project planners can design processes that better meet goals of agencies and the public. Participants in more-intensive processes can improve their outreach and garner more input from the wider public. Agencies can better support participation efforts with adequate staff and resources.

However, the best way to improve public participation is to develop a much broader recognition that its role and goals are central to sound public policy. Many of those responsible for making and implementing environmental policy have granted public participation only grudging acceptance. The managerial model, in which regulatory and administrative decisions are made by appointed experts, has given some ground to pluralist and popular democracy. However, managerialism still holds a powerful grip on many of our decisionmaking processes, especially on technically complex issues.

In the grudging view, public participation is a marginal addition—or even an afterthought—to a fundamentally technical decision process. From this perspective, the most that can be hoped for from members of the public is that they do no harm—that they do not degrade the quality of decisions as measured by risk minimization, economic efficiency, cost-effectiveness, or other technical criteria. Too often, agencies see active citizens and organized communities as opponents and impediments to sound decisions. This unenthusiastic tolerance of a public role easily degenerates into mere public relations whereby decisionmakers attempt to sell their favored outcome to an uninformed public.

Society and the environment would be better served by turning the grudging model on its head. Rather than seeing policy decisions as fundamentally technical with some need for public input, we should see many more decisions as fundamentally public with the need for some technical input. Decisions that engage people's values and interests broadly—as do most of those that make up the case history of public participation—are most appropriately addressed through decision processes that bring those values and interests to the fore.

Turning grudging acceptance on its head will require more than marginal adjustments in technical processes ill-suited to address values and interests. It will require some public role in defining the problems and developing options for solving them, as well as choosing among those options. The history of public participation shows that such a process can make active and organized citizens and communities assets to decisionmaking, rather than perceived impediments.

To enable the public to take on more fundamental roles in decisionmaking, public participation processes do need to effectively incorporate technical information, education, and analysis. Public deliberation and technical analysis can then create a virtuous cycle, with one process adding to the effectiveness and integrity of the other. More-intensive processes—rather than the all-too-common public hearing—are most likely to create such a cycle.

Much of the appeal of technical decisionmaking tools comes from their managerial convenience: they are well understood and well integrated into the practice of environmental professionals. Procedures for participatory decisionmaking are not. We hope that our work can help change this situation.

The foundation for building an understanding of public participation is the case study record—a rich, extensive, and chaotic history of what public participation has accomplished, what has worked, and what has not. The case study record shows us that public participation is more than just a theoretically appealing component of democracy; public participation helps agencies and the public meet concrete challenges that face the modern environmental management system.

The environmental arena has no shortage of challenging decisions ahead. Global environmental problems have risen to the top of the agenda, as have intensely local issues such as urban runoff, water supply, growth management, and transportation. Society's challenge is to make decisions on these issues in ways that educate, reflect public values, resolve conflict, and build trust. Good technical analysis is clearly part of the solution, but only as one component of processes that truly integrate public interaction, public analysis, and public judgement into policy decisions.

Areas for Further Research

By analyzing patterns across cases, we have gleaned some lessons on how to make public participation most effective in the future. But to better understand what public participation can accomplish and identify better processes for getting there, much more research is required.

One important priority for future research is understanding how the context of decisionmaking affects the success of public participation. In our study, attributes of context such as the nature of the issue, the quality of pre-existing relationships, and the institutional setting bore little direct relationship to the success of the process. Yet it is highly unlikely that contextual issues are irrelevant.

Future analyses should consider a wider range of contextual measures detailed in our database. They include problem complexity, the degree of scientific uncertainty, geographic scale, issues related to jurisdictional authority, differences in values among participants, and the degree of potential for win–win solutions. Future research also should examine how contextual issues influence choices about the public participation process (such as the type of mechanism chosen and the motivation of participants) and how process designs make certain contextual issues more or less relevant to success.

Research is also needed to hone our understanding of the relationships among various process attributes. Questions that might be asked include the following:

- How does responsiveness of the lead agency affect the motivation of participants?
- What combinations of process attributes need to be present for success?
- What role does seeking consensus play in enhancing deliberation?

Such research can help identify strategies for enhancing the key determinants of success. Ultimately, we would like to understand how processes evolve over time as demands change, contexts shift, and relationships among parties develop.

Finally, more research on implementation is needed. The value of public participation will ultimately be judged by its ability to enhance implementation and show demonstrable benefits for environmental quality. Understanding the links between participation and actions on the ground is a high priority. Research should focus on the specific links between public participation and the political, legal, and social forces that drive implementation forward. They might include the role of participation in mobilizing and publicizing the public will, the role of participation in building capacity and coordination for implementation, and the connections between participation and external administrative and political processes. This research can draw on an already large body of literature on policy implementation.

Many of these directions for future study—especially in terms of understanding the role of contextual issues and the relationship between participation and implementation—pose a particular challenge for researchers of public participation. Researchers need to broaden the scope of their analysis beyond the participation effort itself to the larger political landscape and historical context in which that participation occurs. Some processes must be observed over time to understand the dynamic relationships among context and process variables as well as how the rationale and goals for participation evolve over time.

Considerations of future research raise questions of methodology. Meta-analysis, such as that used here, is highly useful for identifying broad trends across cases that can be turned into hypotheses for future study. Much information in our case study database remains untapped and can be used in future analyses. However, meta-analysis is limited by the kinds of information that case study authors choose to collect and by its relative blindness to the nuances of individual cases. Research on public participation is probably best served by an iterative research approach in which broad trends drawn from meta-analytical studies such as this one are tested and refined by evaluating carefully chosen case studies. Ultimately, we will be most confident in research results that arise out of (and converge from) various methodological approaches.

Appendix A

Case Survey Methodology

This appendix provides the details of the case survey methodology described in Chapter 2. We describe how we selected cases for analysis, coded data from the case studies, and ensured consistency in coding between researchers.

Case Study Selection

To identify cases appropriate for this study, we searched the literature as far back as the early 1970s. A summer research assistant assisted in reviewing the literature and identifying cases. Using citation databases, publication bibliographies, personal contacts with other researchers, Web-based searches of universities and other research organizations, and our own collections, we identified more than 1,800 publications for review.

Initially, we screened the abstract of each document to identify whether the case in question

- involved public participation,
- occurred in the United States, and
- concerned the environment.

After the initial screening, we brought 531 documents in house for more thorough review. The documents were screened based on whether they described a case that involved

- a discrete mechanism or set of mechanisms—such as public hearings, advisory committees, or environmental mediation—intentionally instituted to engage the public in administrative environmental decisionmaking (this criterion excluded cases concerned mainly with protests, lawsuits, and advocacy campaigns);

- sufficient information on the context, process, and results of the participation effort (this criterion generally excluded cases shorter than five pages long);
- participation of nongovernmental citizens (this criterion excluded several cases of intergovernmental decisionmaking);
- at least one public perspective other than the regulated community, such as an environmental group or community group (this criterion excluded some cases of negotiated settlement of regulatory actions, for example); and
- either an identifiable lead agency or an agency for which the output of the process would be immediately relevant.

After the second screening, 276 documents remained for intensive coding by the researchers. These documents described 333 cases of public participation, because some documents reported multiple cases. (Some cases were covered by multiple documents as well). During the coding process, 128 cases were rejected because in-depth reading revealed that the cases did not meet the screening criteria; these deletions left us with 205 coded cases. To these cases were added 34 cases from a pilot project on public participation in the Great Lakes (Beierle and Konisky 1999), giving a final data set of 239 cases.

The sources of case studies are listed below.

Bibliographic Citation Databases Reviewed

Text searches were performed using combinations of "public or citizen or stakeholder" and "participation or involvement" and "environment or natural resources" and "case or case study" in four databases: Dissertation Abstracts, Enviroline, NTIS, and PAIS International.

Organizations Contacted for Case Studies

Center for Environmental Communication, Rutgers University
Center for Research in Conflict and Negotiation, Pennsylvania State
    University
Colorado Center for Environmental Management
Conflict Research Consortium, University of Colorado
Consensus Building Institute, Inc.
Consortium for Risk Evaluation with Stakeholder Participation
Environmental Conflict Working Group, George Mason University
Human Dimensions Research Unit, Department of Natural Resources, Cornell
    University
Institute for Conflict Analysis and Resolution, George Mason University
Institute for Water Resources, U.S. Army Corps of Engineers

International City/County Management Association
Marasco Newton Group, Ltd.
Meridian Institute
Northeast–Midwest Institute
Pacific Northwest National Laboratory
Policy Consensus Initiative
Program for Community Problem-Solving
RESOLVE, Inc.
U.S. Environmental Protection Agency
Water Environment Federation
Western Rural Development Center

Organizations Whose Publications Lists
Were Examined for Case Studies

Center for Urban and Regional Studies, University of North Carolina
Civic Practices Network
Coalition to Improve Management in State and Local Government
Consortium on Negotiation and Conflict Resolution, Georgia State University
Environmental Council of the States
Kettering Foundation
Lincoln Institute of Land Policy
Mid-Atlantic Center for Community Education
National Academy of Public Administration
National Center for Environmental Decision-Making Research
National League of Cities
National Research Council
Nicholas School of the Environment and Earth Sciences, Duke University
Ohio Commission on Dispute Resolution and Conflict Management
President's Council on Sustainable Development
Program on Negotiation, Harvard Law School
Progressive Policy Institute
RAND Corporation
Risk Assessment and Policy Association
School of Forestry and Environmental Studies, Yale University
School of Natural Resources and Environment, University of Michigan
School of Public and Environmental Affairs, Indiana University
Society for Risk Analysis
Tri-Met
U.S. Institute for Environmental Conflict Resolution
Urban Land Institute
Western Governors' Association
Western Rural Development Center

Conferences Whose Abstracts Were Reviewed

International Symposium on Technology and Society: Technical Expertise
  and Public Decisions (1996)
National Association of Environmental Professionals (NAEP) conference
  proceedings (1998, 1999)
One Day Conference on Risk (1997)
Risk Assessment and Policy Association (1999)
Society for Risk Analysis annual meetings (1986, 1988, 1989, 1994–1998)

Bibliographies and Reference Lists Reviewed

Only the most extensive bibliographies and reference lists are included here.

Bibliographies
Gray and Langton 1995
Jaffray 1981
University of Cincinnati 1998

Reference Lists of Selected Papers
Bauer and Randolph 1999
Berry and others 1984
English, Peretz, and Manderschied 1999
English and others 1993
Lynn and Busenberg 1995
Thomas 1993
Weber 1998
Yosie and Herbst 1998

Case Coding

Cases were coded by the authors (Beierle and Cayford) and a research assistant
(Konisky), using a Microsoft Access database with a custom coding template
on the front end. Each case was coded for more than 100 attributes (see Chapter 2 and Appendix B). Each item received a score (usually high, medium, or
low), a weight-of-evidence designation (also high, medium, or low), and a
descriptive comment. Figure A-1 is a screen shot of a sample template.

In coding the cases, a few unanticipated issues arose, which were resolved
as follows.

A Mix of Participatory Mechanisms

If more than one participatory mechanism was used in a case, we either
focused on the process described in greatest detail or coded the case as a mix-

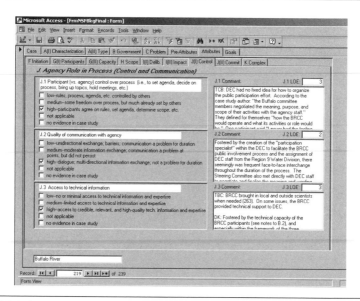

FIGURE A-1. Access Template for Issues Related to Public Control, Communication with the Lead Agency, and Access to Technical Information

ture of processes in the category of the most intensive mechanism (i.e., a case with an advisory committee and a public meeting process would be coded in the advisory committee category).

A Private-Sector Entity Led the Process

Cases were rejected if only a private-sector entity acted as the lead organization. However, if the private-sector entity acted under the requirements of a lead government agency and the results of the process needed to be reviewed and/or approved by a public agency, the case was included. In such cases, the regulatory agency was considered the lead agency and the private-sector party was considered a participant.

Described Public Hearings but Focused Mainly on Conflict External to the Process

Case studies that described public hearings but focused on conflict external to the formal public participation process were rejected because too little information was available about the context, process, and results of the participatory mechanism. To some extent, these kinds of cases can be considered public participation "failures" in that the mechanisms used were not sufficient to resolve controversy, build trust, and so forth. Rejecting them may have biased our pool of cases somewhat toward successes (see Appendix D), but including them would have strengthened the conclusion that less-intensive mechanisms (e.g. public hearings) are often less successful.

Intercoder Reliability

A process of building and testing intercoder reliability was used to ensure that the researchers (Beierle, Cayford, and Konisky) coded cases in the same way. The intercoder reliability process was undertaken in an iterative approach, mainly at the beginning of the project (Yin and Heald 1975). Two researchers coded two to three cases independently, then compared results to identify items that they coded differently. They then developed consensus codes (which were used in the analysis) and made changes to the coding template to reduce ambiguity and confusion. We continued to compare our coding results until we consistently exceeded the two-thirds agreement threshold regarded as satisfactory in the literature (Larsson 1993). From this point, only one researcher coded each case.

Cases for the intercoder reliability process were selected randomly but not through a formalized process (such as random number generation). The cases covered a range of different sources, styles, lengths, environmental issues, and approaches to participation, although they were not explicitly selected to do so. The intercoder reliability process was not just intended as a randomized test of agreement. Rather, it was a process of building consistency among researchers as well as testing for it.

In all, intercoder reliability tests were conducted on 23 cases, or around 10% of the total. The literature provides little guidance on the appropriate number of cases to code for such a test, and we felt that 10% was sufficient. To ensure that researchers continued to code in the same way, intercoder reliability tests were conducted periodically throughout the coding process.

Results of the intercoder reliability tests show the percent agreement after a double-blind coding process for all of the attributes, then for all of the attributes that had been coded with a moderate or high level of evidence (Table A-1). Only the latter data were included in our analysis, and a few attributes that the coders consistently coded differently were excluded.

TABLE A-1. Results of the Intercoder Reliability Tests

	Case	Researchers	Case order in coding sequence	Agree (%) All data	Agree (%) Moderate–high evidence
1	Michigan Relative Risk Analysis Project[a]	Beierle and Konisky	1	62	62
2	Saginaw Bay[a]	Beierle and Konisky	2	51	52
3	Lower Green Bay and Fox River[a]	Beierle and Konisky	3	64	67
4	Rochester Embayment[a]	Beierle and Konisky	4	68	75
5	Minnesota Relative Risk[a]	Beierle and Konisky	5	79	82
6	Citizen Participation at Portland General Electric	Beierle and Konisky	39	84	78
7	Goodyear Tire and Rubber Company	Beierle and Konisky	40	79	74
8	Pine Street Barge Canal Coordinating Council	Beierle and Konisky	41	74	69
9	New Bedford Harbor Forum	Beierle and Konisky	42	74	69
10	Raymark Industries Site (Superfund)	Beierle and Konisky	43	65	59
11	Centroport, USA	Beierle and Konisky	71	86	86
12	Upper Clark Fork River	Beierle and Konisky	72	88	88
13	Indian Ford Creek Mediation	Beierle and Konisky	114	76	75
14	Maryland Oyster Roundtable	Beierle and Cayford	120	71	71
15	Louisiana Black Bear Conservation Plan	Beierle and Cayford	122	77	76
16	Carolawn, Inc., Community Advisory Group	Beierle and Cayford	125	60	59
17	Water Quality Planning Process 208 Program, Erie and Niagara Counties, NY	Beierle and Cayford	136	67	66
18	Environment 2010	Beierle and Cayford	138	76	75
19	Transportation Master Plan for Boulder	Beierle and Cayford	139	64	64
20	Reformulated Gasoline Reg Neg	Beierle and Cayford	141	82	81
21	The Fort Ord Restoration Advisory Board	Beierle and Cayford	140	65	65
22	Public Participation in the CHEAP Proposal	Beierle and Cayford	206	74	74
23	Oregon's Coastal Planning Commission	Beierle and Cayford	207	66	64

[a]Reliability test was conducted during the pilot study.

Appendix B

Details on Data and Aggregation

The public participation case studies were coded for more than 100 attributes. The main attributes that were coded for the categories of context, process, and results and implementation as well as basic case study information are listed in Figures B-1 through B-4, respectively.

In this appendix, we discuss the approach used to aggregate attributes into broad general measures. We sought to identify clusters of attributes that were related both conceptually and statistically. The process of identifying conceptually related attributes was informal and exploratory; it started from our qualitative understanding of the cases. To establish statistical relationships, we considered correlations among variables above 0.45 as "high," correlations between 0.3 and 0.45 as "medium," and correlations below 0.3 as "low."

Aggregation of Context Attributes

The only two context attributes that are aggregated are the measures for preexisting conflict and preexisting trust. In each case, they are aggregates of three different kinds of attributes:

- preexisting conflict or mistrust among the participants (Measure 1),
- preexisting conflict or mistrust in the wider public (Measure 2), and
- indirect measures of preexisting conflict or mistrust (Measure 3).

Indirect measures of preexisting conflict or mistrust are various aspects of the issue(s) that would lend themselves to conflict or mistrust. For preexisting conflict, for example, we examined whether issue resolution raised significant potential conflict between environmental and economic goals, whether the balance of power among stakeholders was historically uneven, whether the

Type of Issue

- Policy-level versus site-specific
- Pollution versus natural resource
- Specific issue categories (facility siting, hazardous waste cleanup, or permitting)
- Planning process versus implementation process*

Preexisting Relationships

- Conflict among public
  - Preexisting conflict among participants ⎫
  - Preexisting conflict in the wider public ⎬
  - Indirect measures of conflict ⎭
- Mistrust of government
  - Preexisting mistrust between participants and agency ⎫
  - Preexisting mistrust between wider public and agency ⎬
  - Indirect measures of mistrust ⎭

Institutional Setting

- Level of government (local, state, federal)
- Identity of lead agency
  - Relationship to problem (manager of problem or cause of problem)*
- Lead agency's level of involvement (directly leading process or not)
- Complexity of agency jurisdictions over problem*
- Whether participation was discretionary *

Other Issues

- Problem complexity*
- Scientific understanding*
- Geographic complexity*
- Location (state, and so forth)*
- Dates*

FIGURE B-1. Context Attributes

Notes: Aggregated variables are identified with braces. Items marked with an asterisk (*) were not considered in this report.

Type of Mechanism and Characteristics

- Type of mechanism (e.g., public meetings and hearings)
- Use of consensus
- Type of output (information, recommendations, or agreement)
- Use of a facilitator
- Duration and frequency*
- Scope of tasks (what the process was meant to accomplish)*
- Ongoing participation versus a finite event*
- Access to technical information*
- Type of participants
  - Participants' access to process (open access versus selection process)
  - Average versus elite
  - Socioeconomic representativeness
- Capacity of participants
  - Political capacity
  - Technical capacity
  - Participatory capacity

Responsiveness of the Lead Agency

- Direct measure of commitment
- Indirect measure of commitment
- Communication between agency and participants
- Agency participation as "partner" in process*

Motivation of the Participants

- Ambition (direct measure)
- Evolution of scope
- Optimism
- Public's commitment to issue*
- Public's perceived influence on process output*
- Public's perceived influence on policy outcomes*

Quality of Deliberation among Participants

- Quality of deliberation among participants

Degree of Public Control

- Bottom up versus top down
- Timing of participation
- Participants' control over process execution
- Who determined scope (participants or agency)?*

Other Features

- Leadership*
- Participants' understanding of their role and the process' goals*
- Internal trust formation (among participants)*
- Perceived fairness of process*

FIGURE B-2. Process Attributes

Notes: Aggregated variables are identified with braces. Items with an asterisk (*) were not considered in this report.

Output

- Goal 1: Incorporating public values into decisions
  - Socioeconomic representativeness
  - Incorporation of wider public values
  - Interest group representation
  - Interest group balance*
- Goal 2: Improving the substantive quality of decisions
  - Direct measures ⎫
  - Indirect measures ⎬

Relationships

- Goal 3: Resolving conflict among competing interests
  - Issues avoided
  - Interests missing
- Goal 4: Building trust in institutions
  - Direct measures ⎫
  - Indirect measures ⎬
  - Increasing trust among wider public

Capacity Building

- Goal 5: Educating and informing the public
  - Wider public education (outreach)
  - Social learning*
- Political influence of participants*
- Creation of an organization to pursue future work*
- Participants' motivation to continue working on issue(s)*

Other Evaluation Information

- Case study author evaluation of case*
- Participants' evaluation of case*
- Evaluation by other institutions*

Implementation

- Stage of implementation
- Likelihood of implementation
- Forces other than public participation influencing implementation

FIGURE B-3. Results and Implementation Attributes

Notes: Aggregated variables are identified with braces. Items with an asterisk (*) were not considered in this report.

Coding Information

- Coder name
- Coding date

Bibliographic Information

- Bibliographic citation(s) of case study or studies

Case Study Author Information

- Affiliation of author(s)
- Research approach(es) (surveys, interviews, and so forth)
- Selected case to illustrate success or failure

FIGURE B-4. Case Study Information Attributes

costs of resolving the problem were highly concentrated on one group of stakeholders, and whether the issue raised high stakes in terms of risk or cost.

The aggregate was developed by using a hierarchical process. If Measure 1 was coded, it was used as the aggregate; if not, Measure 2 was used, and so on. When all measures were scored for a given case, they were generally consistent with each other.

Aggregation of Process Attributes

Capacity of Participants

The capacity aggregate is an average (rounded up) of three individual capacity attributes: political capacity, technical capacity, and participatory capacity. The different measures are highly correlated with each other, and all correlations are significant at a confidence level of $p < 0.01$ (Table B-1). An aggregate was developed even if only one capacity measure was scored.

Motivation of the Participants

This aggregate gauges the ambition and optimism that carries the public forward in a participation process. It combines two measures: ambition, which itself is an aggregate, and optimism.

The ambition aggregate was developed from two attributes: one that coded information on ambition directly, and one that examined whether participants expanded or reduced their scope of work. The decision rules that were used to construct the ambition aggregate are as follows.

1. If the scope of work expanded, then score the case "high"; if the scope of work decreased, then score the case "low."

TABLE B-1. Correlation among Capacity Measures

	Political capacity	*Technical capacity*
Technical capacity	0.62***	
	(*n* = 74)	
Participatory capacity	0.52***	0.46***
	(*n* = 45)	(*n* = 56)

Note: n = number of matched pairs used to calculate the correlation coefficient.
***Significant at *p* < 0.01.

2. If the scope of work did not change or if there is no evidence of change in the scope of work, then use the direct measure for ambition.
3. If the two measures of ambition are contradictory (i.e., high/low combinations), then make case-by-case adjustments.

The optimism attribute, also used to construct the aggregate for motivation of participants, measures why people participate and how much effort they are likely to put into it. The correlation between the ambition aggregate and the optimism attribute is moderate to high (0.44).

The decision rules that were used for developing an aggregate for the overall motivation of participants are as follows.

1. If only one of the two components (ambition or optimism) is scored, use it as the aggregate score (applied to 79 cases).
2. If scores match, use the matching score (applied to 18 cases).
3. If scores are mixed, give low/medium combinations a "low," high/medium combinations a "high," and low/high combinations a "medium" (applied to 17 cases).

Degree of Public Control

Public control measures the extent to which participants (rather than the lead agency) control the initiation, design, and execution of the public participation process. It combines two attributes: public control over process initiation (which itself is an aggregate) and participants' control over process execution.

Public control over process initiation is an aggregate of two variables: whether the process was initiated by an agency or by participants (bottom up versus top down), and the extent to which participation was undertaken early in the decisionmaking process. The aggregate scores were defined as follows:

- low—Participation occurred too late in the process to affect major outcomes, and participants were not involved in its initiation.
- medium—Participants were involved in process execution but not planning, and the process was initiated by the public or by government.

- high—Participants were involved in process planning and execution, and the process was initiated by the public or by government.

The overall public control aggregate combines the aggregate for project initiation with an attribute that rates public control over process execution. The two measures are highly correlated across cases (0.64), signaling that public control over the initiation of a process generally carried over into public control over its execution.

The decision rules that were used for developing the overall aggregate for public control are as follows.

1. If only one of the two components (project initiation aggregate or public control over process execution attribute) is scored, use it as the aggregate score (applied to 70 cases).
2. If scores match, use the matching score (applied to 85 cases).
3. If scores are mixed, give low/medium combinations a "low," medium/high combinations a "high," and low/high combinations a "medium" (applied to 55 cases).

Responsiveness of the Lead Agency

This aggregate rates the interrelated aspects of commitment and communication between the lead agency and the participants. It combines the attributes of communication and commitment, which itself is an aggregate of two attributes: a direct measure of commitment (e.g., adequate resources and staff support) and an indirect measure of commitment (e.g., obvious accountability and the involvement of high-level decisionmakers). The direct measure of commitment was coded as "high," "medium," or "low," and the indirect measure was coded as "yes" or "no." The aggregate commitment measure was simply an average of the scores after converting "yes" scores to "high" and "no" scores to "low." The average was rounded up to "high" for high/medium combinations and to "medium" for medium/low combinations. (The averages of only six cases were rounded.)

The responsiveness aggregate combines the commitment aggregate with an attribute for communication between the lead agency and participants. The two measures are quite highly correlated across cases (0.55).

The decision rules that were used for creating the overall lead agency responsiveness aggregate are as follows.

1. If only one component (commitment aggregate or communication attribute) is scored, use it for the aggregate score (applied to 84 cases).
2. If scores match, use the matching score (applied to 56 cases).
3. If scores are mixed, give low/medium combinations a "low," medium/high combinations a "high," and low/high combinations a "medium" (applied to 26 cases).

Aggregation of Social Goals

Two of the social goals are aggregates: improving the substantive quality of decisions and building trust in institutions. The overall measure of success is an aggregate of all the social goals.

Improving the Substantive Quality of Decisions

The aggregate for this goal was constructed from eight quality criteria (see Chapter 3). The measures can be divided into two groups. Direct measures (cost-effectiveness, joint gains, opinion, and other measures) compare decisions with an implicit baseline and were scored as "high," "medium," or "low." Indirect measures (added information, technical analysis, innovative ideas, and holistic approach) determine what participants add to decision-making and were scored as "yes" or "no."

Because the appropriate criteria varied from case to case, only rarely were more than two criteria scored for each case. In fact, of 172 cases for which at least one of the eight criteria was scored, none was scored on more than five criteria, and only 47 were scored on three to five criteria. Because the quality criteria were so widely distributed (i.e., lacked substantial overlap across cases), intercorrelating them in search of an overarching concept of "quality" was problematic. The largest number of pairwise comparisons that could be made was for 32 cases, and for many pairs of criteria, correlation coefficients could not be calculated because there was no variation in one of the criteria (Table B-2). Nevertheless, some of the eight criteria appear to correlate quite well.

The rules that were used to construct the aggregate are as follows.

1. For an indirect quality measure, convert "yes" scores to "high" and "no" scores to "low."
2. If there are no low/high combinations for a given case (suggesting a wide divergence of scores), average the scores and round up. For cases with low/high combinations, skip to Step 3.
3. If there are low/high combinations, determine scores on a case-by-case basis after reviewing qualitative data.

In developing an aggregate measure for 172 cases, 163 (95%) could be scored after Step 2. For nine cases, the aggregate was determined on a case-by-case basis (Step 3); all of these cases were given "medium" scores on the basis of mixed results.

Building Trust in Institutions

The trust goal is an aggregate of a direct measure of trust and an indirect measure of trust. The indirect measure is itself an aggregate and includes

TABLE B-2. Correlation among Substantive Quality Criteria

	Cost-effective	Joint gains	Opinion	Other direct	Added information	Technical analysis	Innovative ideas
Joint gains	−0.17 (n = 11)						
Opinion	0.49 (n = 7)	0.64** (n = 18)					
Other direct	1.0 (n = 4)	−0.37 (n = 11)	6/8^a agree				
Added information	0.26 (n = 4)	0.60 (n = 13)	6/8^a agree	4/4^a agree			
Technical analysis	0.26 (n = 4)	12/13^a agree	0.82** (n = 10)	4/4^a agree	0.84*** (n = 32)		
Innovative ideas	0.63 (n = 6)	14/18^a agree	0.65 (n = 8)	2/2^a agree	0.85*** (n = 13)	1.0*** (n = 16)	
Holistic approach	−0.13 (n = 6)	7/9^a agree	9/9^a agree	1/2^a agree	1.0** (n = 12)	1.0*** (n = 11)	1.0*** (n = 10)

Note: n = number of matched pairs used to calculate the correlation coefficient.

[a]No coefficient was calculated because of no variation in one of the criteria. The ratio of perfect agreements to total number of pairs is reported.

**Significant at $p < 0.05$ using Fisher's exact test because of the low number of observations.

***Significant at $p < 0.01$ using Fisher's exact test because of the low number of observations.

- increased (or decreased) confidence in agencies' abilities,
- perceptions that an agency would (or not) "do what was right,"
- perceptions that the process was (not) legitimate, and
- other related issues mentioned in case studies.

The individual indirect measures of trust were coded as "yes" or "no." An aggregate indirect measure is an average of all the indirect scores, and ties were rounded to "yes." (Only 5 of 55 cases had to be rounded.) To create an overall aggregate, the "yes" scores were changed to "high" and "no" scores to "low."

To create the aggregate, the direct measure of trust was used where available; where it was not, the aggregated indirect measure was used. In only one case did the direct and indirect scores conflict (i.e., a high/low combination); in that case, the direct measure took precedence.

Aggregate Measure of Success

The aggregate measure of success is a combination of the individual social goal measures. As discussed in Chapter 3, there are several reasons to believe that the five social goals are conceptually related. The goals also are quite well correlated quantitatively (Table B-3).

The highest correlations are between building trust and resolving conflict

TABLE B-3. Correlation among Social Goals

	Educating and informing the public	Incorporating public values into decisions	Improving the substantive quality of decisions	Resolving conflict among competing interests	Building trust in institutions
Incorporating public values into decisions	0.5*** (n = 103)				
Improving the substantive quality of decisions	0.16** (n = 96)	0.36*** (n = 154)			
Resolving conflict among competing interests	0.36*** (n = 89)	0.57*** (n = 149)	0.36*** (n = 137)		
Building trust in institutions	0.47*** (n = 52)	0.46*** (n = 74)	0.46*** (n = 62)	0.57*** (n = 64)	
Aggregate measure of success	0.56*** (n = 104)	0.73*** (n = 160)	0.61*** (n = 151)	0.75*** (n = 146)	0.69*** (n = 73)

Note: n = number of matched pairs used to calculate the correlation coefficient.

** Significant at $p < 0.05$.

*** Significant at $p < 0.01$.

and between resolving conflict and incorporating public values; both relationships have a correlation coefficient of 0.57. In turn, the correlation between trust and all the other goals is greater than 0.46. The correlation between educating and informing the public and incorporating public values in decisions is also high, with a correlation of 0.5. Of the remaining correlations, three are 0.36 and one (between educating and informing the public and improving the substantive quality of decisions) is only 0.16.

To generate an aggregate goal score, the five goal scores were averaged. Ties between medium and high (average aggregate score of 2.5) were rounded up to high, and ties between medium and low (average aggregate score of 1.5) were rounded down to low. Only cases in which three or more goals had been scored received an aggregate score.

Another way to analyze the relationship between individual goal scores is to compare the scores that went into making the aggregate. Of the 166 cases for which an aggregate could be calculated, all the individual goal scores were the same (all high, all medium, or all low) in 42% of the cases. In 37% of cases, the individual goal scores differed from each other by only one category (all high/medium combinations or all medium/low combinations). In only 20% of cases were the high and low individual goal scores mixed.

Appendix C

Technical Analysis

A s much as possible, we exclude details on the statistical analysis of data from the main text of this book and report them in this appendix. In the first section, we describe the bivariate correlations between success and the attributes for context and process. In the second section, we describe the results of a multivariate analysis using an ordered probit regression.

Bivariate Correlations

The bivariate correlations in this analysis use a Kendall's tau b correlation coefficient (Stata 1997), which is an appropriate nonparametric measure of correlation for the kind of ordinal data derived in the coding process (Bullock and Tubbs 1987). The coefficient is calculated based on the number of concordant and discordant pairs of observations in a contingency table, using a correction for ties. The contingency table can be represented as a simple table in which the values of one variable are row headings and the values of a second variable are column headings. In each cell of the table, n (the number of cases in which the two variables take on the corresponding row and column values) is recorded.

The statistical significance of the correlations is determined using a χ^2 test. A rule of thumb for using the χ^2 test is that the expected n of each cell in the contingency table should be greater than 5, preferably greater than 10 (Stokes, Davis, and Koch 1995). This criterion was met in most correlations described here.

All of the bivariate correlations referred to in the main text of the book are listed in Table C-1. The relationships of those correlated variables are discussed in detail in the main text, so only the main points are mentioned here. First, the correlations between success and the context attributes (type of

TABLE C-1. Correlations between Success and Other Case Attributes

Attribute	Correlation with success
Context	
Type of issue	
Policy level vs. site specific	0.02
	($n = 166$)
Pollution vs. natural resource	–0.01
	($n = 166$)
Preexisting relationships	
Conflict among public[a]	0.13**
	($n = 145$)
Mistrust of government[a]	0.05
	($n = 96$)
Institutional setting	
Level of government	–0.07
	($n = 158$)
Level of involvement of lead agency	0.04
	($n = 166$)
Process	
Type of mechanism	
Intensity of the participatory mechanism	0.43***
	($n = 166$)
Variable process features	
Responsiveness of the lead agency[b]	0.56***
	($n = 81$)
Motivation of the participants[b]	0.44***
	($n = 53$)
Quality of deliberation[b]	0.39***
	($n = 61$)
Degree of public control[b]	0.25*
	($n = 84$)
Implementation	
Likelihood of implementation (Stage 3)	0.40***
	($n = 52$)
Likelihood of implementation (Stage 4)	0.35**
	($n = 64$)

Note: n = number of matched pairs used to calculate the correlation coefficient.

[a] For both preexisting conflict and preexisting mistrust, low = significant preexisting conflict or mistrust exists, whereas high = little preexisting conflict or mistrust. A positive correlation, then, means that better preexisting relationships are more conducive to public participation success.

[b] As discussed in Chapter 5, correlations are for a truncated data set that excludes cases of public meetings and hearings as well as cases in which the lead agency was not directly involved.

* Significant at $p < 0.10$.   ** Significant at $p < 0.05$.   *** Significant at $p < 0.01$.

issue, preexisting relationships, and institutional setting) are all small and, with one exception, not statistically significant. The only statistically significant result is a small positive relationship between preexisting conflict and successful public participation. This correlation suggests a weak relationship between less initial conflict and better participation.

Second, the correlation between success and the intensity of the participatory mechanism is moderate to high and is highly statistically significant. As mechanism intensity increases, so does the likelihood that public participation will be successful. The success of advisory committees that use consensus and those that do not differs little. Accordingly, if intensity of the participatory mechanism is compressed to only three categories (public hearings and meetings, advisory committees, and negotiations and mediations), then the correlation rises to 0.47 and remains significant at $p < 0.01$.

Third, the correlations between success and the four variable process features range from high to low. Responsiveness of the lead agency is most related to success and degree of public control is least related to success; motivation of the participants and the quality of deliberation fall between the two extremes. For all of the process features except degree of public control, the correlations are highly statistically significant ($p < 0.01$). Finally, the correlations between success and the two measures of implementation are moderate but statistically significant. The correlation decreases slightly from changes in laws, regulations, and policy (Stage 3) to actions taken on the ground (Stage 4).

Multivariate Data Analysis

A multivariate ordered probit regression model is used to examine how independent variables (context attributes and process attributes) explain variation in a dependent variable (the aggregate measure of success). The advantage of such a model over bivariate correlations is its ability to examine the influence of many independent variables on a dependent variable. For example, it allows us to examine how type of mechanism influences success while controlling for the influence of issue type, level of government, and whatever other variables are introduced into the analysis.

The ordered probit model is one of a group of models of qualitative choice (Pindyck and Rubinfeld 1991). Such models are common in the analysis of survey data. They are used when the value taken on by the dependent variable (in this case, the aggregate measure of success) is measured in terms of ordered categories (such as our measures of high, medium, and low). The probit model assumes that the dependent variable has an underlying normal distribution, but we can see that distribution only by looking at the three categories. The probit model represents the dependent variable as a *cumulative normal probability function*; that is, the model estimates the likelihood or probability that a public participation process will be successful. Because a normal distribution is assumed, standard tests of statistical significance can be used.

Use of this model assumes that the likelihood of success is based on a linear combination of a series of independent variables (in this case, the context and process attributes). All of the independent variables used here are dummy variables that take on a value of either 0 or 1. Generally, a positive (and statistically significant) coefficient on a dummy variable means that it increases the likelihood of a successful process. A negative (and statistically significant) coefficient on a dummy variable means that it reduces the likelihood of success.

When several dummy variables are used to represent a series of exclusive categories, one of the dummies always must be dropped in the regression analysis to avoid perfect collinearity. For example, when we analyze six environmental issues (design of permits and operating requirements, facility siting, and so forth), we introduce only five dummy variables into the analysis. The sixth alternative is represented when all of the five dummy variables take on values of 0, so introducing a sixth variable would be redundant. When a dummy variable is dropped in this way, the coefficients on the remaining variables are interpreted as effects relative to the dropped variable. For example, if we drop the dummy variable for cases that deal with natural resources planning and management, the interpretation of the coefficient on the dummy variable for facility siting cases describes how changing the issue under discussion from natural resources planning and management to facility siting (all else equal) affects success.

In the regression, the dependent variable SUCCESS is the aggregate measure of success. It takes a value of 1 for cases that received low scores, 2 for cases that received medium scores, and 3 for cases that received high scores.

The first set of independent variables describes the context of participation. Each type of issue is represented by a dummy variable that takes a value of 1 for cases dealing with that type of issue and 0 otherwise. They include REGULATION, for cases that involve the design of regulations and standard setting; SITING, for cases of facility siting; HAZWASTE, for cases that entail investigation and cleanup of hazardous waste sites; PERMITTING, for cases that include the design of permits and operating requirements; RESOURCES, for cases of natural resources planning and management; and POLICY, for cases regarding policy development and comparative risk assessment. In the multivariate regression, the variable RESOURCES is dropped to avoid perfect collinearity.

The variable RELATIONSHIP records the quality of preexisting relationships. It takes a value of 1 if there is preexisting conflict or mistrust. (Note that the sign on the RELATIONSHIP variable should be interpreted differently than the signs for preexisting conflict and preexisting mistrust in the bivariate analysis. In the multivariate analysis, a negative coefficient on RELATIONSHIP means that better preexisting relationships are more conducive to success).

The level of government is recorded as FEDERAL for cases that involved federal agencies, STATE_LOCAL for cases that involved state and local agencies, or OTHER for cases that involved other kinds of lead agencies. Each of these variables takes a value of 1 if it is true in a particular case and 0 otherwise. In the multivariate regression, STATE_LOCAL is dropped to avoid perfect collinearity.

To account for cases in which the lead agency is not directly involved in the process, AGENCY_INVOLVE is included in the analysis. It takes a value of 1 if the agency is directly involved and 0 otherwise.

The four main kinds of public participation processes are recorded as MEETING for public hearings and meetings, COMM_NOCON for advisory committees not seeking consensus; COMM_CON for advisory committees seeking consensus, and NEGOTIATE for negotiations and mediations. Each of these variables takes a value of 1 in cases where they are true and 0 otherwise. To avoid perfect collinearity, MEETING is dropped in the multivariate regression.

The specific process features of responsiveness of the lead agency, motivation of the participants, quality of deliberation, and degree of public control are represented by a single aggregate measure. The aggregate is used to avoid complications from data gaps as well as to account for the moderate degree of intercorrelation among the four process measures. The aggregate is simply a rounded average of the four scores for all cases that were scored on three or more of the process features. High/medium ties are rounded up to "high," and medium/low ties are rounded down to "low." Cases that received high scores on the aggregate take a value of 1 for the variable PROCESS_H and 0 otherwise; cases that received medium scores take a value of 1 for the variable PROCESS_M and 0 otherwise; cases that received low scores take a value of 1 for the variable PROCESS_L and 0 otherwise. To avoid perfect collinearity, PROCESS_L is dropped in the multivariate regression.

Of the 239 cases in the data set, 142 cases contain data for all of the variables of interest. The distribution of these cases across types of issue, levels of government, mechanism types, and other variables is roughly similar to the data set as a whole. The sample means and standard deviations for each of the included variables and the results of the multivariate data analysis are listed in Table C-2.

Also reported in Table C-2 are the results from two models. Model 1 includes variables for three categories of context attributes (type of issue, preexisting relationships, and institutional setting) and one process category (type of mechanism). Within each category are appropriate dummy variables. Model 2 is the same as Model 1 plus an additional category for the variable process features.

TABLE C-2. Results of Multivariate Analysis (Dependent Variable = SUCCESS)

	Sample means (SD)	Model 1 Coefficients (SE)	Model 2 Coefficients (SE)
Context			
Type of issue[a]			
REGULATION	0.07	0.25	1.14
	(0.26)	(0.73)	(1.02)
SITING	0.14	0.20	–0.16
	(0.35)	(0.37)	(0.41)
HAZWASTE	0.20	0.03	0.03
	(0.40)	(0.34)	(0.36)
PERMITTING	0.10	0.39	0.51
	(0.30)	(0.49)	(0.55)
POLICY	0.18	–0.60*	–0.71*
	(0.38)	(0.34)	(0.38)
Preexisting relationships			
RELATIONSHIP	0.63	–0.39	–0.18
	(0.49)	(0.25)	(0.28)
Institutional setting[b]			
FEDERAL	0.43	–0.59**	–0.52*
	(0.50)	(0.26)	(0.28)
OTHER	0.05	–0.19	–0.02
	(0.22)	(0.65)	(0.86)
AGENCY_INVOLVE	0.72	0.15	–0.25
	(0.45)	(0.27)	(0.31)
Process			
Type of mechanism[c]			
COMM_NOCON	0.27	1.05***	0.28
	(0.45)	(0.36)	(0.41)
COMM_CON	0.30	1.50***	0.56
	(0.46)	(0.35)	(0.41)
NEGOTIATE	0.27	2.66***	1.56***
	(0.45)	(0.48)	(0.56)
Variable process features[d]			
PROCESS_M	0.30	NA	1.46***
	(0.46)		(0.36)
PROCESS_H	0.53	NA	2.52***
	(0.50)		(0.40)
No. of observations (*n*)	142	142	142
Probability > χ^2		0.00	0.00
Pseudo R^2		0.23	0.42

Notes: Numbers in parentheses indicate standard deviation (SD) or standard error (SE). NA = not analyzed.

[a]Omitted variable = RESOURCES.   [b]Omitted variable = STATE_LOCAL.

[c]Omitted variable = MEETING.   [d]Omitted variable = PROCESS_L.

*Significant at $p < 0.10$.   **Significant at $p < 0.05$.   ***Significant at $p < 0.01$.

In both of the models, many of the process attributes receive high, positive, and statistically significant coefficients, indicating that they have the largest influence on the likelihood of a process being successful. The context attributes demonstrate few relationships with success one way or the other.

A comparison of Models 1 and 2 sheds some light on the relative importance of type of mechanism and the variable process features. Adding the variable process features to Model 2 does three things. First, it shows that good scores on the process features increase the likelihood of success, as demonstrated by relatively large, positive, and statistically significant coefficients. Second, it reduces the magnitude of the coefficient on the type of mechanism variables (making two of them not statistically significant). The decrease in the type of mechanism coefficients suggests that some of the influence demonstrated by the process features in Model 2 is being accounted for by the mechanism types in Model 1. Indeed, the intercorrelation between the intensity of the mechanism types and higher scores on the process features is moderate. Finally, the addition of the process features to Model 2 nearly doubles the explanatory power of the model. The "pseudo" R^2 is 0.23 in Model 1 and rises to 0.42 in Model 2. The pseudo R^2 in an ordered probit model, like the standard R^2 in a linear regression model, measures how much of the variation in the dependent variable is being explained by the model. For a complex social process such as public participation, an R^2 of 0.42 is quite respectable.

The principal lesson from the multivariate analysis is that process is a much better predictor of success than context is. However, coefficients are significant in two context variables. These coefficients are not as large as they are on the process variables, but they deserve some discussion.

The first context variable to note is POLICY. In both Models 1 and 2, the coefficient on POLICY is negative and significant at $p < 0.10$. It suggests that, all else equal, policy development cases are less likely to produce successful participation than are resource planning and management cases (the omitted variable). In fact, even if we use any type of issue as the omitted variable, the coefficient on POLICY continues to be negative and significant. Policy development cases appear to perform consistently worse than other kinds of cases. The result emerges in a multivariate analysis only because of the interaction between policy type and type of mechanism. Policy development cases are among the worse performers in our data set, but not obviously different from cases of facility siting or the investigation and cleanup of hazardous waste sites (Chapter 4, Figure 4-1). What sets policy development cases apart is that they typically use more-intensive mechanisms than do the siting and hazardous waste cases (Chapter 4, Figure 4-2).

Indeed, the statistical result for POLICY shows what a closer inspection of the data reveals: quite a few unsuccessful, or only moderately successful, policy development cases used consensus-based advisory committees. Additional research will have to determine whether the result is an artifact of the data pool used here or whether small-group consensus processes are not well suited to broad policy development exercises in general.

The second context variable with a significant coefficient in both models is FEDERAL. Its negative sign suggests that participation led by federal agencies is less likely to be successful than participation led by state or local agencies (the omitted variable). It is notable that the relationship between level of government and success does not appear in the bivariate analysis. The poorer performance of federal-led cases is apparent only when other factors are controlled for. This result is interesting because one would expect federal-led processes to have more resources and more political clout than their state and local counterparts. Instead, the state and local cases perform better—a good sign, if trends toward devolution and local decisionmaking continue.

Appendix D

Examination of Potential Bias

In this appendix, we examine whether the case study data set is biased. In particular, we are interested in whether various influences may have made the pool of case studies we collected and analyzed for our study appear more successful than the public participation norm. We discuss the issue of bias generally, examine six possible sources of bias qualitatively, then examine the two most likely sources of bias quantitatively.

Our overall conclusion from the examination of bias is that the case study pool may be moderately biased toward successful cases; however, that bias does not change our main findings. Even when we account for potential bias, the pool of case studies still appears to be much more successful than not, and process is still much more related to success than is context.

Only in the analysis of implementation does bias appear to affect important results of our study. Accounting for bias reduces the evidence of a link between good participation and good implementation. This finding strengthens our argument that implementation is influenced by many factors in addition to the quality of the public participation efforts that precede it.

General Discussion of Bias

Bias can be a problem for any kind of communication, and the kinds of communication we relied on in performing this study are no exception. Consider the process by which we collected data on public participation cases:

1. An agency chooses to use a public participation process.
2. A case study author decides to write about a particular public participation process.

3. A case study author chooses study methods, data interpretation methods, and the emphasis and tone of the narrative.
4. We identify case studies in our literature review.
5. We use a screening process to select case studies for analysis.
6. We code the attributes of each case.

Any of these steps could have introduced bias. We could have been misled by a biased author who could have been misled by biased participants who provided data for the case study. We could even be misleading *you*, the reader. There are also many hidden influences: somehow, the case study author's work got published and came to our attention; somehow, this book was published and came to your attention. Every step of the process that requires communication is potentially biased.

The solution to the problem of possible bias is not a perfect method that yields absolute truth. Besides being impossible, such an ideal method would not remove the bias that is reintroduced with every communication. Suppose that we could detect and correct for every distortion communicated by the participants to the authors and by the authors to us. We could then know that our results were perfect. However, you still would not know whether our results are correct, because you would not know whether we had applied that ideal method correctly and reported the results faithfully.

People do many things to improve the reliability of their methods, including bringing in eyewitnesses to testify, inviting observers from both parties to monitor counts, and repeating experiments with different scientists. All these are worthwhile efforts to get closer to an ideal method. We do discuss the use of such techniques to improve the reliability of our methods in the remainder of this appendix, but such techniques were not our primary approach to minimizing bias.

We minimized bias primarily by taking the widest view to date of the historical record of public participation. We used as many cases of public participation as we could find. We selected cases from every kind of source that we could find, and we used multiple sources for the same cases whenever possible. Having read hundreds of public participation case studies, we can report that some of them are biased. Sometimes the bias was blatant, and other times it was revealed in other sources. We did not remove such papers from our database, though we noted their bias. To remove such papers might have decreased author bias but also would have increased our influence on the case study data set. We minimized bias not by trying to ferret out and eliminate each occurrence of it—a highly unreliable strategy—but by taking a bigger picture of public participation than individual biases could possibly distort.

That said, we turn to some obvious ways in which bias may have entered the data set and consider how they might affect the analysis.

Possible Sources of Bias

From the real world to our data set, bias may have been introduced into the case studies during any of the six important steps listed above.

Step 1: An Agency Chooses to Use a Public Participation Process

Most environmental decisions are not made through public participation. In those that are, participation is either required or discretionary. In cases where participation is required, there would be no obvious bias toward participation being successful or unsuccessful.

In cases where participation is discretionary (i.e., most cases in our data set), one can imagine bias working toward greater success or greater failure. On the one hand, agencies may be motivated to institute public participation processes only when the likelihood of success is high. For example, they may want to increase the likelihood of a successful pilot or demonstration project, or they may want to avoid instituting a process that will exacerbate existing conflict. On the other hand, public participation processes may be initiated precisely because participation is the only alternative for dealing with highly controversial or highly complex decisions. In such cases, participation would be, a priori, more likely to fail.

We suggest that agencies' choices to use public participation processes are probably neutral in terms of biasing the case study pool toward success.

Step 2: A Case Study Author Decides to Write about a Particular Public Participation Process

Case study authors may be motivated to write about successful cases, just as scientists report successful experiments. If this were true, the case study record might demonstrate a "selection bias" toward more successful cases. However, several reasons lead us to believe that selection bias may not exist or, if it exists, may not bias our cases toward success.

The first issue is timing. Many authors selected cases for analysis in the early stages, before success or failure would have been apparent. By the time processes were completed, authors might have invested years of work in direct observation and other data collection, creating a strong motivation to publish the work regardless of the success of the process.

The second issue is the tendency toward comparative case study research. Many documents, particularly doctoral dissertations, discuss and compare several cases. Reviewing both successful and unsuccessful cases makes it easier for the authors of these case studies to draw conclusions and illustrate their points.

The third issue is multiple definitions of success. Even if authors were motivated to write about only successful cases, they used many definitions of success. Some authors considered cases successful if participants reached consensus, others considered any fair process successful, and still others considered progress toward implementation successful. If an author's notion of success was largely independent of the social goals by which we evaluate cases, then any bias from the author would not affect our results.

The final issue is ambivalence about public participation. Not all authors who write about public participation are fans of it. Some may want to emphasize the failings of these kinds of processes. Selection bias in the case study record, then, could be toward success or failure of the process.

In examining possible selection bias, we do not have to rely on only conceptual arguments. During the coding process, researchers recorded information on possible selection bias, which we analyze quantitatively at the end of this appendix.

Step 3: A Case Study Author Chooses Study Methods, Data Interpretation Methods, and the Emphasis and Tone of the Narrative

After choosing to write about a case, an author's methods, mode of interpretation, and personal motivations may drive him or her to emphasize its successes rather than its failures. We call this phenomenon "author bias." It is most likely to occur when the case study author is closely connected to the process (e.g., as a participant, a facilitator, or a lead agency liaison) rather than an independent researcher.

The affiliation(s) of the case study author(s), which we use as an indicator of potential author bias, was noted during the coding process, and we use the information to examine bias quantitatively at the end of this appendix.

Step 4: We Identify Case Studies in Our Literature Review

There is little reason to believe that the methods we used to identify cases would result in a biased sample. Case identification involved a comprehensive search of published articles and books, doctoral dissertations and master's theses, government and institutional reports, conference proceedings, discussion papers, self-published papers, and draft manuscripts.

Although one might expect some bias in a subset of these sources, a consistent bias across the entire data set is unlikely.

Step 5: We Use a Screening Process to Select Case Studies for Analysis

The criteria we used to screen cases for inclusion in the study (see Appendix A) may have biased the pool of case studies in some way, although the screening criteria had nothing to do directly with the success or failure of cases. In fact, a good balance between successes and failures would have been very desirable.

One screening criterion in particular may have had an indirect influence on the success of the pool of cases. Cases that dealt with citizen activism were not included in our study because they did not focus on "a discrete mechanism (or set of mechanisms) intentionally instituted to engage the public in administrative environmental decisionmaking." The backdrop to these cases was often a set of public hearings or public comment periods required by administrative procedure rules. When such participatory efforts failed to satisfy the public, the alternative was protest.

Cases rejected from our sample because they focused on activism could be interpreted as cases of failed public hearings and public comment. Their exclusion would bias the case study pool toward success, but only for public hearings and meetings. Accounting for this possible source of success bias, then, would only emphasize the conclusion drawn throughout this book: that less-intensive participatory mechanisms were less successful.

Step 6: We Code the Attributes of Each Case

In coding the cases, the researchers sought an unbiased representation of what they contained. Self-interest would encourage balanced reporting, to ensure both the integrity of the data (to avoid having to code the cases again!) and maximum variation in the data on which to perform the analysis. As suggested by Yin and Heald (1975), information on data quality was recorded so that poorly supported data could be eliminated from the analysis. The intercoder reliability process also helped avoid individual coders' possible biases.

One possible source of bias that had little to do with our intentions was how to treat data gaps. This issue arose mainly when the coding called for "yes" or "no" answers to questions such as "Did participants add useful information to the decision process?" If the case study had positive or negative information in this regard, how to code the response was obvious. However, when the case study had no information, interpreting whether the item should be left blank or coded as "no" was sometimes difficult. If the item was left blank, the assumption was that the case study author simply did not focus on that particular issue. If it was coded as "no," the assumption was that the case study author was reporting only what the participants did, rather than what they failed to do.

We chose to leave such items blank rather than interpret the absence of data as "no." However, depending on how these data gaps are interpreted, they may create some success bias. Unfortunately, knowing for sure whether such a bias exists is impossible. We suggest that there may be a small success bias for the two goals that depended the most on such questions: improving the substantive quality of decisions, and building trust in institutions.

Quantitative Analysis of Bias

Two of the most likely sources of bias discussed above can be analyzed quantitatively. *Selection bias* resulted when authors chose to write about only successful cases. Of 239 cases, 66 cases (27%) appeared to be chosen mainly to illustrate success or failure (however defined) and 114 cases (48%) were not; in 59 cases (25%), data were not available.

Author bias resulted from an incentive to emphasize the successes of a case rather than its failures. Although we cannot know fully the intentions of the authors, we posit that those closely associated with the process would be more likely to want to make the results look good. Of the 239 case studies, 70 cases (29%) were written by someone closely affiliated with the participation process and 150 cases (63%) were not; data were not available for 19 cases (8%).

Combining selection and author bias, we identified 118 (49%) potentially biased cases, 87 (36%) unbiased cases, and 34 (14%) cases for which data were not available. (We should note that the terms "potentially biased" and "unbiased" refer only to the two potential sources of bias examined here, not to all possible sources of bias).

Throughout the rest of this appendix, we examine quantitatively whether consideration of bias leads to conclusions different from those discussed in the main body of the book. We use three approaches:

- comparison of the success of the unbiased and potentially biased sets of cases in achieving the social goals and implementing decisions,
- comparison of the bivariate correlations between success and the context and process attributes for the unbiased and potentially biased sets of cases, and
- multivariate analysis that incorporates variables for possible bias.

First, it is important to note two ways in which the unbiased and potentially biased sets of cases differ. The most important difference has to do with different proportions of mechanism types in the two sets of cases. The pool of unbiased cases contains fewer more-intensive participatory mechanisms than the pool of potentially biased cases does (Figure D-1). The reason that fewer more-intensive mechanisms made it into the unbiased set has much to do with the relative ease of identifying the "success" of more-intensive mechanisms. Many cases were deemed successful if consensus was reached; then, many such cases were selected and case studies written to demonstrate how to pursue consensus successfully, raising questions of selection bias. Additionally, the more-intensive mechanisms were usually facilitated, and facilitators often authored the case studies themselves, raising questions of author bias.

The unbiased cases differed from the overall pool of cases in another important way that becomes apparent in the multivariate analysis: the unbiased set contains no cases of regulatory design.

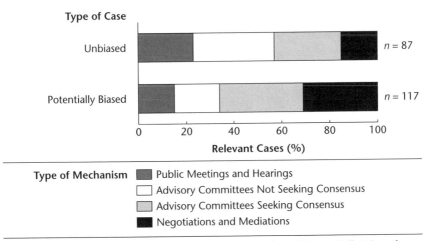

FIGURE D-1. Type of Mechanism Used in Unbiased and Potentially Biased Cases

Note: n = total number of cases scored.

Success and Implementation of Unbiased and Potentially Biased Cases

Unbiased cases were somewhat less successful than potentially biased cases (Figure D-2). However, most of the difference in levels of success can be attributed to the concentration of less-intensive mechanisms among the unbiased cases. More-intensive mechanisms were found to be more successful (but cases describing them were also more likely to be biased), so fewer of these more-intensive mechanisms necessarily lead to a less successful group of cases.

Even though the unbiased cases appear to be less successful than the potentially biased cases, they still compose quite a successful group. More than 40% of the unbiased cases were highly successful, and fewer than 20% were unsuccessful. Our general conclusion about the impressive success of public participation efforts still holds.

The difference between unbiased cases and potentially biased cases is more dramatic for some of the implementation data (Figure D-3). The cases in the unbiased set were much less likely to be implemented than those in the potentially biased set, at least in terms of Stage 3 (changes in law, regulation, or policy). This quantitative result corroborates our impression that many authors were overly optimistic about the likelihood of implementation (discussed in Chapter 6). They either picked cases that illustrated successful implementation or were led to overemphasize the probable success of implementation because of their closeness to the participation effort.

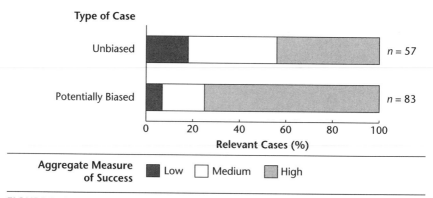

FIGURE D-2. Aggregate Measure of Success for Unbiased and Potentially Biased Cases

Note: n = total number of cases scored.

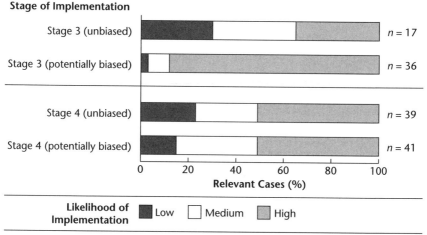

FIGURE D-3. Likelihood of Implementation for Unbiased and Potentially Biased Cases

Notes: n = total number of cases scored. Stage 3 is "changes in law, regulation, or policy"; Stage 4 is "actions taken on the ground."

Bivariate Correlations

The correlations between success and the context and process attributes for the unbiased and the potentially biased sets of cases are compared in Table D-1. The main results are generally the same as those for the entire set of cases described in Appendix C. In both sets of cases, most of the context attributes remain uncorrelated with success, whereas the process features are still highly correlated. However, a few differences emerge.

TABLE D-1. Correlations between Success and Other Case Attributes for Unbiased and Potentially Biased Cases

Attribute	Correlation with success	
	Unbiased	Potentially biased
Context		
Type of issue		
Policy level vs. site specific	0.06	–0.12
	(*n* = 57)	(*n* = 83)
Pollution vs. natural resource	–0.12	–0.06
	(*n* = 57)	(*n* = 83)
Preexisting relationships		
Conflict among public[a]	0.32**	0.12
	(*n* = 51)	(*n* = 71)
Mistrust of government[a]	0.13	0.26
	(*n* = 34)	(*n* = 48)
Institutional setting		
Level of government	–0.10	–0.02
	(*n* = 54)	(*n* = 78)
Level of involvement of lead agency	0.09	0.19
	(*n* = 57)	(*n* = 83)
Process		
Type of mechanism		
Intensity of mechanism	0.34	0.36**
	(*n* = 57)	(*n* = 83)
Variable process features		
Responsiveness of the lead agency[b]	0.52**	0.55***
	(*n* = 29)	(*n* = 43)
Motivation of the participants[b]	0.51*	0.54***
	(*n* = 18)	(*n* = 28)
Quality of deliberation[b]	0.55***	0.49***
	(*n* = 22)	(*n* = 34)
Degree of public control[b]	0.23	0.16
	(*n* = 29)	(*n* = 44)
Implementation		
Likelihood of implementation (stage 3)	0.35	0.20
	(*n* = 13)	(*n* = 34)
Likelihood of implementation (stage 4)	0.27	0.53**
	(*n* = 27)	(*n* = 28)

Note: n = number of matched pairs used to calculate the correlation coefficient.

[a]For both preexisting conflict and preexisting mistrust, "low" scores indicate that significant preexisting conflict or mistrust existed, whereas "high" scores indicate that there was little preexisting conflict or mistrust. A positive correlation, then, means that better preexisting relationships are more conducive to public participation success.

[b]As discussed in Chapter 5, correlations are for a truncated data set that excludes cases of public meetings and hearings and cases where the lead agency was not directly involved.

*Significant at *p* < 0.10.   **Significant at *p* < 0.05.   ***Significant at *p* < 0.01.

First, in the analysis of unbiased cases, preexisting conflict is moderately positively correlated with success, and it is statistically significant. (The positive correlation here means that good relationships among parties—less preexisting conflict—lead to more successful cases.) For the potentially biased cases, the correlation is small and not significant. Again, the difference can probably be explained by differences in type of mechanism. The set of potentially biased cases contains more more-intensive mechanisms than the set of unbiased cases does. Because the more-intensive mechanisms do a better job of transforming preexisting conflict than less-intensive mechanisms do, the correlation between preexisting conflict and the ultimate success of the process is weaker in the potentially biased set with its higher proportion of more-intensive mechanisms.

Second, the correlation between success and the intensity of the mechanism is not statistically significant for the unbiased cases but is statistically significant for the potentially biased cases. The difference in significance is probably mostly due to the number of cases for which we have observations (57 for unbiased cases and 83 for potentially biased cases). However, it also may indicate that we overstated the relationship between success and the intensity of the participatory mechanism. Although this finding may temper the conclusion that success and type of mechanism are related (Chapter 5), it does not affect the conclusion that the variable process features are highly related to success. The high, positive, and statistically significant results for responsiveness of the lead agency, motivation of participants, and quality of deliberation in the unbiased and potentially biased sets of cases still emphasize the importance of process.

Finally, the correlations between successful participation and the likelihood of implementation are not statistically significant for the unbiased cases. The high and significant correlation between success and implementation (Stage 4) for the potentially biased set of cases only emphasizes that the link between participation and implementation is probably an artifact of bias.

Multivariate Analysis

A multivariate analysis allows us to examine the impact of potential bias simply by introducing dummy variables for bias into the regression equation. Doing so lets us observe how bias affects the results for other variables in the model and compare the results with the data presented in Appendix C. Using the multivariate analysis also helps us determine whether bias can explain some of the likelihood of success across cases.

We used three kinds of models (Table D-2). In Models 1 and 2, we introduce the single dummy variable BIAS. Its value is 1 if there is potential selection bias or author bias for the case and 0 otherwise. We have data on BIAS for

TABLE D-2. Results of Multivariate Analysis Controlling for Potential Bias (Dependent Variable = SUCCESS)

Attribute	Sample mean (SD)	Model 1 (SE)	Model 2 (SE)	Model 3 (SE)
Context				
Type of issue[a]				
SITING	0.16	0.08	–0.56	–0.76
	(0.37)	(0.41)	(0.47)	(0.68)
HAZWASTE	0.19	0.04	0.07	–0.89
	(0.39)	(0.36)	(0.40)	(0.57)
PERMITTING	0.11	0.40	0.58	0.74
	(0.31)	(0.53)	(0.61)	(0.87)
POLICY	0.18	–0.79**	–1.0**	–1.8***
	(0.39)	(0.37)	(0.43)	(0.58)
Preexisting relationships				
RELATIONSHIP	0.62	–0.59**	–0.45	–0.45
	(0.49)	(0.29)	(0.32)	(0.39)
Institutional factors[b]				
FEDERAL	0.41	–0.57**	–0.51*	–0.78*
	(0.49)	(0.28)	(0.31)	(0.42)
OTHER	0.06	–0.29	0.02	–0.37
	(0.23)	(0.66)	(0.93)	(1.08)
AGENCY_INVOLVE	0.72	0.20	–0.41	–0.86*
	(0.45)	(0.31)	(0.36)	(0.51)
Process				
Type of mechanism[c]				
COMM_NOCON	0.26	0.84**	–0.38	–0.28
	(0.44)	(0.40)	(0.50)	(0.72)
COMM_CON	0.32	1.3***	–0.09	–0.16
	(0.47)	(0.40)	(0.50)	(0.76)
NEGOTIATE	0.26	2.3***	0.95	0.12
	(0.44)	(0.52)	(0.64)	(0.85)
Variable process features[d]				
PROCESS_M	0.31	NA	1.81***	1.65***
	(0.46)		(0.43)	(0.60)
PROCESS_H	0.52	NA	2.79***	2.56***
	(0.50)		(0.47)	(0.64)
Potential bias				
BIAS	0.57	0.57**	0.60**	NA
	(0.50)	(0.27)	(0.30)	
BIAS_A	NA	NA	NA	0.99*
				(0.55)
No. of observations (*n*)	121	121	121	65
Probability > χ^2		0.000	0.000	0.000
Pseudo R^2		0.23	0.43	0.36

Notes: Numbers in parentheses indicate standard deviation (SD) or standard error (SE). NA = not analyzed.

[a]Omitted variable = RESOURCES and REGULATION.　　[b]Omitted variable = STATE_LOCAL.

[c]Omitted variable = MEETING.　　[d]Omitted variable = PROCESS_L.

*Significant at $p < 0.10$.　　**Significant at $p < 0.05$.　　***Significant at $p < 0.01$.

121 of the 142 cases examined in Appendix C. Of the 121 cases, 52 cases (43%) are unbiased, and 69 cases (57%) are potentially biased. Except for the introduction of the BIAS variable, Models 1 and 2 are the same as the multivariate models in Appendix C. The mean values of each variable demonstrate that the 121 cases examined are roughly similar to the 142 cases examined in Appendix C. However, as noted above, the unbiased sample contains no regulatory design cases, so we dropped the variable for REGULATION to avoid overdetermination.

Model 3 takes a different approach to examining bias. All cases for which selection bias is evident are eliminated, and only the role of author bias is examined. The dummy variable for author bias, BIAS_A, takes a value of 1 if the case has author bias and 0 if not.

Model 3 examines whether treating selection bias and author bias differently affects the results. Whereas selection bias affects whether the case enters our initial pool of data (i.e., whether it is written up as a case study), author bias affects the scores of various attributes. Therefore, it may be most appropriate to simply exclude the cases that entered our data set as a result of selection bias but to examine how author bias affects results for the remaining set of cases. In this model, we have 65 case studies, 52 (80%) of which are unbiased and 13 (20%) of which are potentially biased.

Regardless of which model we look at, introducing variables for bias into the multivariate analysis does not change significantly the conclusions presented in Appendix C. As before, the importance of type of mechanism dominates the regression in Model 1 and is supplanted in Models 2 and 3 by the variable process features. Again, policy development cases and those led by the federal government appear to be less likely to succeed than other cases.

The only real difference that results from introducing potential bias into the model is that the coefficient on preexisting relationships is significant in Model 1. The negative sign on the coefficient should be interpreted the same way that a positive sign is interpreted in the bivariate analysis: better preexisting relationships lead to more successful public participation. As in the bivariate analysis, the significant result probably can be explained by the higher proportion of less-intensive mechanisms in the unbiased set of cases than in the potentially biased set of cases.

Results of all the models indicate that the variable for potential bias explains some of the success found in the cases. The coefficients on the bias variables are positive and statistically significant in all three models. However, these coefficients are not large relative to other significant process variables. Introducing bias variables into the analysis also does not increase the pseudo R^2 considerably over the models shown in Appendix C. Success bias plays a role, but not a prominent one.

Appendix E

Citations for Cases in the Final Data Set

Facility Siting Cases

Brooklyn Navy Yard Incinerator (Ozawa 1991a, 1991b)

CAG for Maine Low Level Radioactive Waste Authority (Clary and Hornney 1995)

California Power Plant Siting (Lake 1980b)

Central Arizona Water Control Study (Brown 1984)

Centroport, USA (Mazmanian and Nienaber 1979)

Chem-Nuclear Siting Public Participation (Webler 1992)

Citizen Participation at Portland General Electric (Mogen 1986)

Coal Creek Project (McConnon 1986)

Colorado Springs Water Alternatives (Cortese and Firth 1997)

DuPont Collaborative Process in Okefenokee (*Atlanta Journal-Constitution* 1999; *Georgia Times-Union* 1999; RESOLVE 1999)

Energy Facility Siting—East Bend, Kentucky (Whitlach and Aldrich 1980)

Energy Facility Siting—Mountaineer, West Virginia (Whitlatch and Aldrich 1980)

Flood Control on the Snoqualmie (Mazmanian and Nienaber 1979)

Guilford County Task Force (Aronoff and Gunter 1994; Lynn 1987)

Interstate 90 (Cormick and Patton 1980; Talbot 1983)

Johnson Creek Case (Seltzer 1983)

Missouri River Basin Plan (Mazmanian and Nienaber 1979)

Northern States Power Advisory Task Force I (Ducsik and Austin 1986)

Northern States Power Advisory Task Force II (Ducsik and Austin 1986)

Orange County Landfill Siting Process (Miranda, Miller, and Jacobs 1996)

Port Townsend Terminal (Talbot 1983)

Siting a Limestone Mine on Laurel Mountain (Steelman and Carmin 1998)

Siting a New County Landfill at DR-7 (McComas and Scherer 1998)

Siting an Energy Facility in Maryland (McConnon 1986)

Snoqualmie River Mediation (Bacow and Wheeler 1984; Cormick and Patton 1980; Mernitz 1980)

Southern Water Supply Project (Linscott 1994)

Tanner Act—Buttonwillow Local Assessment (Cole 1999)

Tanner Act—Kings County Local Assessment (Cole 1999; Menton 1996; Press 1994)

Tanner Act—Martinez Local Assessment (Cole 1999; Menton 1996)

Texas High-Speed Rail EIS (Miller and Griffith 1994)

Three Rivers Watershed (Mazmanian and Nienaber 1979)

WATERS Projects, Glendale, Arizona (Alberhasky and Rozelle 1996)

West Side Highway Mediation (Bacow and Wheeler 1984; Lake 1980a)

White Salmon River Power Development (Howell, Olsen, and Olsen 1987)

Whitney Mine Controversy (Valdez 1993)

Wildcat and San Pablo Creeks (Mazmanian and Nienaber 1979)

Hazardous Waste Cases

Ashtabula River (Ashtabula River Partnership website: http:/www.epa.gov/ region5/arp0.htm; Hartig and Law 1994; Landre 1991; Landre and Knuth 1993; Letterhos 1992; Ohio EPA 1991; project website: http:// www.epa.gov/glnpo/aoc/ashtabula.html)

Black River (project website: http://www.epa.gov/glnpo/aoc/blackriver.html; Hartig and Law 1994; IJC 1996, 1998)

Brio Oil Refinery Community Assistance Panel (Cole and Stevens 1996)

Brio Refinery Community Advisory Group (U.S. EPA 1996; Cole and Stevens 1996)

Buffalo River (Hartig and Law 1994; IJC 1996; Kellogg 1993b)

Carolawn, Inc. Community Advisory Group (U.S. EPA 1996)

Citizen Radiation Monitoring Program (Gray 1989; Gricar and Baratta 1983)

Clinton River (Hartig and Law 1994; *Michigan Area of Concern News* 1992; SPAC 1997)

Colorado School of Mines Research Institute CAG (U.S. EPA 1996)

Cuyahoga River (Becker 1996; Beeker, Studen, and Stumpe 1991; Hartig and Dolan 1995; Hartig and Law 1993, 1994; IJC 1996; project website: http://www.epa.gov/glnpo/aoc/cuyahoga.html)

Detroit River (Becker 1996; Canada Centre for Inland Waters website: http://www.cciw.ca/glimr/raps/connecting/detroit/intro.html; Carpenter 1997; Kellogg 1993a; Landre and Knuth 1990)

Fernald Environmental Management Project (Applegate 1998; Duffield and Depoe 1997)

Fort Ord Restoration Advisory Board (Houghton and Siegel 1997; Szaz and Meuser 1995; Wernstedt and Hersh 1997)

Goodyear Tire and Rubber Company (Susskind, Podziba, and Babbitt 1989)

Grand Calumet River and Indiana Harbor Ship (CARE 1997; Gould 1991a, 1991b; Hartig and Law 1994; Holowaty and others 1992; IJC 1998; Landre and Knuth 1990; project website: http://www.epa.gov/glnpo/aoc/grandcal.html)

Hanford Site Specific Advisory Board (Bradbury and Branch 1999; U.S. DOE 1997)

Hazardous Waste Landfill (Anonymous DOE site) (Apostolakis and Pickett 1998)

Idaho National Engineering and Environment Lab CAB (Bradbury and Branch 1999; U.S. DOE 1997)

Lipari Landfill Superfund Site (Kaminstein 1996; Kauffman 1992)

Lower Green Bay and Fox River (Becker 1996; Gurtner-Zimmermann 1994, 1996; Hartig and Dolan 1995; Hartig and Law 1994; IJC 1996; Kraft and Johnson 1998; MacKenzie 1993, 1996; project website: http://www.epa.gov/glnpo/aoc/greenbay.html)

Manistique River (Gould 1991a, 1991b; Gould, Schnaiberg, and Weinberg 1996; Hartig and Law 1994; IJC 1998)

Massachusetts Military Reservation (Scher 1997)

Menominee River (Hartig and Law 1994; IJC 1996; Landre 1991; Landre and Knuth 1993; SPAC 1997)

Milwaukee Estuary (Hartig and Law 1994; Kaemmerer and others 1992; project website: http://www.epa.gov/glnpo/aoc/milwaukee.html)

Moffett Naval Air Station Superfund site (Di Santo 1998)

Nevada Test Site CAB (Bradbury and Branch 1999; U.S. DOE 1997)

New Bedford (early effort) (Finney and Polk 1995; Schattle 1998)

New Bedford Harbor Forum (Finney and Polk 1995; Hartley 1998, 1999; Schattle 1998)

New York Niagara (Hartig and Law 1994; Kellogg 1993b; project website: http://www.epa.gov/glnpo/aoc/niagara.html)

Northern New Mexico Community Advisory Board (Bradbury and Branch 1999; U.S. DOE 1997)

Nyanza Hazardous Waste Site (Powell 1988)

Oak Ridge Reservation Environmental Management SSAB (Bradbury and Branch 1999; Cusick 1995; U.S. DOE 1997)

Oronogo-Duenweg Mining Belt Site Community Advisory Group (U.S. EPA 1996)

Paducah Gaseous Diffusion Plant SSAB (Bradbury and Branch 1999; U.S. DOE 1997)

Pantex Plant Citizens' Advisory Board (Bradbury and Branch 1999; U.S. DOE 1997)

Pine Street Barge Canal Coordinating Council (Hartley 1998, 1999)

Pollution Abatement Services (Clean Sites 1992)

Raymark Industries Site (Superfund) (Cole and Stevens 1996)

River Raisin (Hartig and Law 1994)

Rochester Embayment (Hartig and Dolan 1995; Hartig and Law 1994; Hartig and Zarull 1992; IJC 1996; Kellogg 1993b; project website: http://www.epa.gov/glnpo/aoc/rochester.html)

Rocky Flats Citizen Advisory Board (Bradbury and Branch 1999; U.S. DOE 1997; Rogers, Murakami, and Hanson 1995; Wilson 1993)

Rouge River (Cole-Misch and Kirschner 1995; Hartig and Dolan 1995; Hartig and Law 1994; Hartig, Thomas, and Iwachewski 1996; IJC 1996; Landre and Knuth 1990; *Michigan Area of Concern News* 1992; Michigan Department of Natural Resources 1994; Schrameck, Fields, and Synk 1992)

Saginaw Bay (Hartig and Law 1994; MacKenzie 1993, 1996; project website: http://www.epa.gov/ecoplaces/part2/region5/site12.html)

Savannah River Site Citizens Advisory Board (Bradbury and Branch 1999; U.S. DOE 1997; Greenberg and others 1997; Patterson, Smith, and Martin 1998)

Sheboygan River & Harbor (Cole and Stevens 1996)

Southern Maryland Wood Treatment Site CAG (Stephan 1998; U.S. EPA 1996)

St. Louis FUSRAP Sites (Simon 1999)

St. Louis River/Bay (Hartig and Dolan 1995; Hartig and Law 1994; IJC 1996; Landre and Knuth 1990; project website: http://www.epa.gov/glnpo/aoc/stlouis.html)

Torch Lake (Hartig and Law 1994; *Michigan Area of Concern News* 1992, 1997; project website: http://www.ss.mtu.edu/EP/Torchlake/AOC_Text.html; SPAC 1997; Torch Lake Public Action Council website: http://www.portup.com/torch/torch1.html)

Tucson International Airport Superfund Site (Di Santo 1998)

Unexploded Chemical Munitions PIRP (Shepherd and Bowler 1997)

Waukegan Harbor (Cole-Misch and Kirschner 1995; Community Advisory Group 1994; Gould 1991a, 1991b; Hartig and Law 1994; IJC 1995, 1996; Michaud 1998; project website: http://www.epa.gov/glnpo/aoc/waukegan.html; Ross, Burnett, and Davis 1992)

Winfield Locks and Dam, Kanawha River (Langton 1996)

Permitting/Operating Cases

Anitec Image Corporation Pollution Controversy (Johnston 1999)

ASARCO–Tacoma Smelter (Call 1985; Kalikow 1984; Krimsky and Plough 1988; Scott 1988; Sirianni 1999)

California Oil Spill Technical Advisory (Busenberg 1997)

Cook Inlet Regional Citizens Advisory Council (Busenberg 1997)

Fitchburg Water Supply Mediation (Edgar 1990a)
General Permit for El Paso and Teller Counties (Lefkoff 1995)
General Permit for Hydrocarbon Development (Delli Priscoli 1988)
General Permit for Wetlands Fill (Delli Priscoli 1988; Rosener 1983)
Harry S. Truman Dam and Reservoir Mediation (Moore 1991)
Hudson River Settlement (Talbot 1983)
Lake Catamount (Schneider 1994)
Maine Oil Spill Advisory Committee (Busenberg 1997)
Pig's Eye Mediation (St. Paul, Minnesota) (Nelson 1990c)
Prince William Sound Regional Citizens Advisory (Busenberg 1997, 1999)
Prince William Sound Risk Assessment (Busenberg 1997, 1999)
Project XL: Intel (Bethell 1997; Freeman 1997; NAPA 1997; Orenstein 1998)
Project XL: Merck (Bauer and Randolph 1999; Orenstein 1998)
Project XL: Weyerhauser (Bethell 1997; NAPA 1997; Orenstein 1998)
Rowley Anti-Snob Zoning Controversy (Lampe and Kaplan 1999; Susskind
    and others 1999)
Sand Lakes Quiet Area Issue-Based Negotiation (Nelson 1990b)
SE Florida Workshops—U.S. Army Corps of Engineers (Rosener 1983)
Sugarbush Water Withdrawal Mediation (Fitzhugh and Dozier 1996)
Swan Lake Conflict (Bingham, Wolf, and Wohlgenant 1994; Talbot 1983)

Policy Development Cases

Athens County Comparative Risk (Institute for Local Government
    Administration and Rural Development 1995; Ohio Comparative Risk
    Project 1995b)
Baltimore Community Environmental Partnership (U.S. EPA 2000)
Bonneville Power Administration (Johnson 1993)
California Utilities Collaborative (English and others 1994)
Chemical Demilitarization Program Case (Bradbury and others 1994;
    Shepherd and Bowler 1997)
Columbus Priorities '95 (Columbus Health Department 1995; NCCR 1996a,
    1996b; Ohio Comparative Risk Project 1995b; Priorities '95 Steering
    Committee 1995; WCED 1997)
Conservation and Load Management Task Force (English and others 1994)
Dakota County Land Use (Jefferson Center 1997)
Dayton Power and Light Company Collaborative (English and others 1994)
Denver's Clean Air Task Force (Stewart, Dennis, and Ely 1984)
Enterprise for the Environment (Clarke 1998; Coglianese 1999a; Enterprise
    for the Environment 1997; Greer 1998; Ruckelshaus 1998)
Environment 2010 (Minard, Jones, and Paterson 1993; Paterson and Andrews
    1995)

Georgia Collaborative (English and others 1994)

Great Lakes Water Levels Study (Edgett 1996, 1997)

Hamilton County Comparative Risk (HCEPP 1998a, 1998b; HCEPP website: http://www.queencity.com/hcepp/; Kaufman 1998)

Idaho Wilderness Controversy (Baird, Maughan, and Nilson 1995)

Illinois Common Ground Consensus Project (Nelson 1990a)

Inland Northwest Field Burning Summit (Mangerich and Luton 1995)

LA/OMA Project (McNelly 1982)

Lower Columbia River Estuary Comparative Risk (Marriott n.d.)

Michigan Relative Risk Analysis Project (Michigan RRAP 1992; MRRAP website: www.deq.state.mi.us/osep/errp.htm; Minard, Jones, and Paterson 1993)

Minnesota Agriculture and Water Quality (Crosby, Kelly, and Schaefer 1986; Jefferson Center 1985)

Minnesota Electricity (Hoffman and Matisone 1997)

Minnesota Relative Risk (Jefferson Center 1996a; project website: http://www.usinternet.com/users/jcenter/jcsumart.html; Schmiechen, Kolze, and Melander 1997)

Minnesota Traffic Congestion Pricing (Jefferson Center 1995)

Moose Management in the Adirondacks (Lauber and Knuth 1996)

National Coal Policy Project (Gray 1989; Gray and Hay 1986)

New Orleans Collaborative (English and others 1994)

Northeast Ohio Comparative Risk Project (Beach 1993; NCCR 1995; Ohio Comparative Risk Project 1995b; WCED 1997)

Ohio Comparative Risk Project (Ohio Comparative Risk Project 1995a, 1995b, 1997; WCED 1997)

Patuxent River Cleanup Agreement (McGlennon 1983)

Promised Land State Park (Gray and Purdy 1992; Purdy and Gray 1994)

Public Meetings on Wildlife Program Management (Siemer and Decker 1990)

Public Participation in the CHEAP Proposal (Swearingen 1998)

Public Service Company of Colorado Collaborative (English and others 1994)

Puget Power Collaborative (English and others 1994)

Tennessee Valley Authority—Deferring Nuclear Projects (Ford 1986)

Virginia Toxics Roundtable (Cook 1983)

Western Mass Electric Company Collaborative (English and others 1994)

Wisconsin Groundwater Legislation Negotiations (Edgar 1990b)

Regulation and Standard-Setting Cases

Asbestos in Schools Reg-Neg (*BNA ADR Report* 1987a, 1987b; Steinzor and Strauss 1987)

Colorado Black Bear Hunting Controversy (Loker and Decker 1994)

Disinfectant By-products Reg-Neg (Bingham, Wolf, and Wohlgenant 1994; Stern and Fineberg 1996)

Emergency Exemptions/Pesticide-Licensing Regulatory Negotiation (Susskind and Van Dam 1986)

Equipment Leaks Negotiated Rule Making (Freeman 1997)

Fall Protection Negotiated Rule Making (Freeman 1997)

Maine's Transportation Policy Reg-Neg (Bogdonoff 1995)

MDA Reg-Neg (Perritt 1995)

Minnesota Hog Farming (Jefferson Center 1996b)

Reformulated Gasoline Reg-Neg (Bauer and Randolph 1999; Weber 1998; Weber and Khademian 1997; Pritzker and Dalton 1995)

Timber/Fish/Wildlife Agreement (Flynn 1992)

Truck-Engine Manufacturers' Reg-Neg (Fiorino and Kirtz 1985; Perritt 1986; Susskind and McMahon 1985)

Woodburning Stove Emissions Reg-Neg (Funk 1987; Ozawa 1991b)

Resource Planning and Management Cases

Animas River Stakeholder Group (Kenney 1997)

Area Guide for White and Green Mountain NFS (Graves and LaPage 1977)

Aviation Corridor Study (Cornelio and Grimm 1982)

Balcones Canyonlands Conservation Plan (Beatley, Fries, and Braun 1995; Bidwell 1998a)

Bighorn River Management Planning Process (McMullin and Nielsen 1991)

Blackfoot Challenge Project (Sullivan 1997)

Bolsa Chica Collaborative Planning (Salvesen 1995)

Boston Water Supply Citizens Advisory Committee (Platt 1995)

Chama River Reservoir (Moore 1997)

Chiwaukee Prairie (Haygood 1995)

Citizen's Task Force on Deer Management (Stout and Knuth 1994, 1995)

Clark County Habitat Conservation Plan (Bernazzani 1998)

Clear Creek Watershed Forum (Kenney 1997)

Columbia River Gorge (Euler 1996)

Conservation & Development in South Walton (Johnson and others 1996)

Countryside Stewardship Exchange (Gross 1996)

Critical Areas Ordinance (Cvetkovich and Earle 1994)

Deer Management in Iowa County (Cavaye 1997)

Devil's Lake State Park CAC (Cavaye 1997)

Durham Citizens' Advisory Committee/CIC (Lynn 1987)

East Everglades Planning Study (Abrams and others 1995)

Ecosystem Management—Tonasket/Okanagon (Geisler and others 1994)

EPA's 208 Water Quality Planning Process: Site A (Plumlee, Starling, and Kramer 1985)

EPA's 208 Water Quality Planning Process: Site B (Plumlee, Starling, and Kramer 1985)

Florida Keys National Marine Sanctuary (Stancik 1995)

Frontlanders Project (Sullivan 1997)

Henry's Fork Watershed Council (Brown 1999; Johnson 1998; Sherlock 1996)

Horicon Marsh Area Coalition (Cavaye 1997)

Indian Ford Creek Mediation (Lampe and Kaplan 1999)

Inimim Forest Management Plan (Duane 1997)

Karner's Blue Butterfly Habitat Conservation Plan (Crismon 1998b)

Local Emergency Planning Committee (Durham, NC) (Amaral and others 1991)

Louisiana Black Bear Conservation Plan (Merrick 1998a)

Maine's Atlantic Salmon Conservation Plan (Opperman 1998a)

Malpai Borderlands Group (Bernard and Young 1997)

Maryland Oyster Roundtable (Chesapeake Bay) (Arnold 1996)

Mattole Watershed Alliance (Bernard and Young 1997)

McKenzie Watershed Council (Kenney 1997)

Metropolitan Seattle Transit Planning Study (Onibokun and Curry 1976)

Missouri Flat Creek Watershed Plan (Osterman and others 1989)

Model Watershed Project (Kenney 1997)

Monongahela National Forest Public Comment (Steelman 1996; Steelman and Ascher 1997)

Northern Forest Lands Council (Ault 1997)

Ohio River Major Investment Study (Lampe and Kaplan 1999; Susskind and others 1999)

Oregon's Coastal Planning Commission (Doubleday, Godwin, and Orange 1977)

Plan for Yosemite National Park (Buck 1984; Buck and Stone 1981)

Pleasant Valley Habitat Conservation Plan (Opperman 1998b)

Plum Creek Habitat Conservation Plan (Miller 1998b)

Poudre Valley Greenbelt Association (Alden 1982)

Quincy Library Group Proposal (Bernard and Young 1997; Duane 1997; Hetherington and Piotrowski 1996)

Red-Cockaded Woodpecker Safe Harbor (Georgia) (Crismon 1998a)

Red-Cockaded Woodpecker Safe Harbor (Texas) (Bidwell 1998b)

Rio Puerco Watershed Committee (Kenney 1997)

Safe Drinking Water Act Compliance Mediation (Clean Sites 1992; Crocker and others 1996)

San Diego Multiple Species Conservation Plan (Merrick 1998b)

San Juan National Forest Mediation (Tableman 1990)

San Pedro River Collaborative Resource Management (Moote, McClaran, and Chickering 1997)

Sandhills Safe Harbor Plan (Miller 1998a)

Santa Fe Summit (Lampe and Kaplan 1999)

Santa Rosa Long-Term Wastewater Project (Marks 1999)

Siskiyou National Forest (Davis 1997)

Statewide 208 Water Quality Planning (North Carolina) (Godschalk and Stiffel 1981)

Stephen's Kangaroo Rat Habitat Conservation Plan (Bernazzani and Opperman 1998)

Transportation Master Plan for Boulder (Kathlene and Martin 1991)

Tulare County Habitat Conservation Plan (Merrick 1998c)

Umatilla Basin Project Mediation (Neuman 1996)

Upper Carson River Basin (Kenney 1997)

Upper Clark Fork River (John and Mlay 1997)

Upper Narragansett Bay (Burroughs 1999)

Verde River Watershed Association (Kenney 1997)

Virginia Coast Sustainable Development Task Force (Bernard and Young 1997)

Water Quality Planning Process 208 Program, Erie and Niagara Counties, NY (Cohen 1979)

Winnebago Comprehensive Management Plan (Cavaye 1997)

Zuni Indian Reservation (Bird 1983)

References

Abrams, K.S., H. Gladwin, M.J. Mathews, and B.C. McCabe. 1995. The East Everglades Planning Study. In *Collaborative Planning for Wetlands and Wildlife,* edited by D.R. Porter and D.A. Salvesen. Washington, DC: Island Press.

Alberhasky, J.E., and M. Rozelle. 1996. Public Involvement: Evolving to Meet a Community's Needs. *Interact* Fall 2(2): 21–33.

Alden, H.R. 1982. Citizen Involvement in Gravel Pit Reclamation: A Case Study. In *Wildlife Values of Gravel Pits: Symposium Proceedings*, edited by W.D. Svedarsky and R.D. Crawford. St. Paul, MN: University of Minnesota Agricultural Experiment Station.

Amaral, D., R. Hetes, F. Lynn, and D. Austin. 1991. Community-Level Use of Risk Analysis: A Case Study. In *Risk Analysis: Prospects and Opportunities*, edited by C. Zervos. New York: Plenum Press.

Anderson, J., and S. Yaffee. 1998. Balancing Public Trust and Private Interest: Public Participation in Habitat Conservation Planning. Summary Report. Ann Arbor, MI: University of Michigan, School of Natural Resources and Environment.

Apostolakis, G.E., and S.E. Pickett. 1998. Deliberation: Integrating Analytical Results into Environmental Decisions Involving Multiple Stakeholders. *Risk Analysis* 18(5): 621–634.

Applegate, J.S. 1998. Beyond the Usual Suspects: The Use of Citizens Advisory Boards in Environmental Decisionmaking. *Indiana Law Journal* 73(3): 1–43.

Arnold, A. 1996. Maryland Oyster Roundtable: Beyond the Distributive Barrier. *RESOLVE* 27.

Arnstein, S.R. 1969. Ladder of Citizen Participation. *Journal of the American Institution of Planners* 35: 216–224.

Aronoff, M., and V. Gunter. 1994. A Pound of Cure: Facilitating Participatory Processes in Technological Hazard Disputes. *Society and Natural Resources* 7(3): 235–252.

Ashford, N.A. 1984. Advisory Committees in OSHA and EPA: Their Use in Regulatory Decisionmaking. *Science, Technology, and Human Values* 9: 72–82.

Atlanta Journal-Constitution. 1999. DuPont Aborts Mining Project. February 6.

Ault, K.E. 1997. The Value of Narratives in Public Hearings: Observed Functions beyond the Formal Objectives in Two Northern Forest Lands Council Listening Sessions in the Adirondacks (New York). M.S. thesis. Syracuse, NY: State University of New York College of Environmental Science and Forestry.

Bacow, L., and M. Wheeler. 1984. *Environmental Dispute Resolution*. New York: Plenum Press.

Baird, D., R. Maughan, and D. Nilson. 1995. Mediating the Idaho Wilderness Controversy. In *Mediating Environmental Conflicts*, edited by J.W. Blackburn. Westport, CT: Quorum, Chapter 16, 229–245.

Bauer, M.J., and P.J. Randolph. 1999. Improving Environmental Decision-Making through Collaborative Methods. *Policy Studies Review* 16(3/4): 168–191.

Beach, D. 1993. Environmental Priorities: What Problems Do We Tackle First? How Do We Decide? *Ecocity Cleveland* 1(6).

Beatley, T., T.J. Fries, and D. Braun. 1995. The Balcones Canyonlands Conservation Plan: A Regional, Multi-Species Approach. In *Collaborative Planning for Wetlands and Wildlife*, edited by D.R. Porter and D.A. Salvesen. Washington, DC: Island Press.

Becker, M. 1996. Implementing a Binational Ecosystem Management Strategy in the Great Lakes Basin: Will the Remedial Action Policy Process Succeed in Restoring the Areas of Concern? Ph.D. dissertation. Durham, NC: Duke University, Department of Environmental Studies.

Beeker, J., G. Studen, and L. Stumpe. 1991. The Cuyahoga Remedial Action Plan Coordinating Committee: A Model for Building Community Ownership of a Watershed Restoration Plan. In *Surface and Ground Water Quality: Pollution Prevention, Remediation, and the Great Lakes*, edited by A.A. Jennings and N.E. Spangenberg. Bethesda, MD: American Water Resources Association.

Beierle, T.C. 1999. Using Social Goals to Evaluate Public Participation in Environmental Decisions. *Policy Studies Review* 16(3/4): 75–103.

———. 2000. The Quality of Stakeholder-Based Decisions: Lessons from the Case Study Record. Discussion paper 00–56. Washington, DC: Resources for the Future.

———. 2001. *Democracy On-Line: An Evaluation of the National Dialogue on Public Involvement in EPA Decisions*. Report. Washington, DC: Resources for the Future.

Beierle, T.C., and D. Konisky. 1999. Public Participation in Environmental Planning in the Great Lakes Region. Discussion paper 99–50. Washington, DC: Resources for the Future.

———. 2000. Values, Conflict, and Trust in Participatory Environmental Planning. *Journal of Policy Analysis and Management* 19(4): 587–602.

———. 2001. What Are We Gaining from Stakeholder Involvement? Observations from Environmental Planning in the Great Lakes. *Environment and Planning C: Government and Policy* 19(4): 515–527.

Bernard, T., and J. Young. 1997. *The Ecology of Hope*. Gabriola Island, British Columbia, Canada: New Society Publishers.

Bernazzani, P. 1998. Clark County Habitat Conservation Plan. In *Improving Integrated Natural Resource Planning: Habitat Conservation Plans*. National Center for Environmental Decision-Making Research website. http://ncedr.org/casestudies/hcp.html (accessed October 14, 2001).

Bernazzani, P., and J. Opperman. 1998. Riverside County Stephen's Kangaroo Rat Habitat Conservation Plan. In *Improving Integrated Natural Resource Planning: Habitat Conservation Plans*. National Center for Environmental Decision-Making Research website. http://ncedr.org/casestudies/hcp.html (accessed October 14, 2001).

Berry, J., K. Portney, M.B. Bablitch, and R. Mahoney. 1984. Public Involvement in Administration: The Structural Determinants of Effective Citizen Participation. *Journal of Voluntary Action Research* 13: 7–23.

Bethell, C. 1997. Learning by Doing: Project XL's Midcourse Correction. *Corporate Environmental Strategy* 4(Summer): 14–22.

Bidwell, D. 1998a. Balcones Canyonlands Conservation Plan. In *Improving Integrated Natural Resource Planning: Habitat Conservation Plans*. National Center for Environmental Decision-Making Research website. http://ncedr.org/casestudies/hcp.html (accessed October 14, 2001).

———. 1998b. Texas Red-Cockaded Woodpecker Safe Harbor. In *Improving Integrated Natural Resource Planning: Habitat Conservation Plans*. National Center for Environmental Decision-Making Research website. http://ncedr.org/casestudies/hcp.html (accessed October 14, 2001).

Bingham, G. 1986. *Resolving Environmental Disputes: A Decade of Experience*. Washington, DC: The Conservation Foundation.

Bingham, G., A. Wolf, and T. Wohlgenant. 1994. *Resolving Water Disputes: Conflict and Cooperation in the United States, the Near East, and Asia*. Arlington, VA: Integrated Irrigation Management Resources.

Bird, M. 1983. A Social Assessment on the Zuni Indian Reservation. In *Public Involvement and Social Impact Assessment*, edited by M. Garcia and J. Delli Priscoli. Boulder, CO: Westview Press, 241–252.

Blahna, D.J., and S. Yonts-Shepard. 1989. Public Involvement in Resource Planning: Toward Bridging the Gap Between Policy and Implementation. *Society and Natural Resources* 2(3): 209–227.

BNA ADR Report. 1987a. EPA Negotiates Proposed Rule on Asbestos in Schools. *BNA ADR Report* 1(July 9): 133.

———. 1987b. Participants and Facilitators Discuss Negotiation of EPA's Proposed Rule on Asbestos in Schools. *BNA ADR Report* 1(July 23): 154.

Bogdonoff, S. 1995. Consensus Building to Write Environmentally Responsive Rules for Maine's New Transportation Policy. In *Mediating Environmental Conflicts*, edited by J.W. Blackburn. Westport, CT: Quorum, chapter 11, 151–166.

Bradbury, J.A., and K.M. Branch. 1999. *An Evaluation of the Effectiveness of Local Site-Specific Advisory Boards for U.S. Department of Energy Environmental Restoration Programs*. Washington, DC: Battelle Pacific Northwest National Laboratory.

Bradbury, J.A., K.M. Branch, J.H. Heerwagen, and E.B. Liebow. 1994. *Community Viewpoints of the Chemical Stockpile Disposal Program*. Washington, DC: Prepared for Science Applications International Corporation by Battelle Pacific Northwest Laboratories.

Breyer, S. 1995. *Breaking the Vicious Circle: Toward Effective Risk Regulation*. Cambridge, MA: Harvard University Press.

Brown, C.A. 1984. The Central Arizona Water Control Study: A Case for Multiobjective Planning and Public Involvement. *Water Resources Bulletin* 20(3): 331–337.

Brown, J. 1999. *On the Henry's Fork: Building a Trust Account*. Ashton, ID: Henry's Fork Foundation.

Buck, J.V. 1984. The Impact of Citizen Participation Programs and Policy Decisions on Participants' Opinions. *Western Political Quarterly* 37(3): 468–482.

Buck, J.V., and B.S. Stone. 1981. Citizen Involvement in Federal Planning: Myth and Reality. *Journal of Applied Behavioral Science* 17(4): 550–565.

Bullock, R.J., and M.E. Tubbs. 1987. The Case Meta-Analysis Method for OD. *Research in Organizational Change and Development* 1: 171–228.

Burroughs, R. 1999. When Stakeholders Choose: Process, Knowledge, and Motivation in Water Quality Decisions. *Society and Natural Resources* 12(8): 797–810.

Busenberg, G.J. 1997. Citizen Advisory Councils and Environmental Management in the Marine Oil Trade. Technical report. Chapel Hill, NC: University of North Carolina, Department of Environmental Sciences and Engineering, Environmental Management and Policy Program, 77.

———. 1999. Collaborative and Adversarial Analysis in Environmental Policy. *Policy Sciences* 32(1): 1–12.

Call, G. 1985. Arsenic, ASARCO, and EPA: Cost-Benefit Analysis, Public Participation, and Polluter Games in the Regulation of Hazardous Air Pollutants. *Ecology Law Quarterly* 12(3): 567–617.

CARE (Citizens Advisory for the Remediation of the Environment). 1997. *Remedial Action Plan Stage II, Grand Calumet Area of Concern.*

Carpenter, E. 1997. *Assessment of the Concerns Related to the Detroit River Binational Public Advisory Committee.* Chicago, IL: Clean Sites.

Cavaye, J.M. 1997. The Role of Public Agencies in Helping Rural Communities Build Social Capital: Case Studies of the Interactions between the Wisconsin Department of Natural Resources and Local Communities. Ph.D. dissertation. Madison, WI: University of Wisconsin–Madison.

Clarke, D. 1998. What Went Right. *The Environmental Forum* March/April: 39–41.

Clary, B.B., and R. Hornney. 1995. Evaluating ADR as an Approach to Citizen Participation in Siting a Low-Level Nuclear Waste Facility. In *Mediating Environmental Conflicts,* edited by J.W. Blackburn. Westport, CT: Quorum, Chapter 9, 121–137.

Clean Sites. 1992. *Superfund Enforcement Mediation: Case Studies.* Chicago, IL: Clean Sites.

Coglianese, C. 1997. Assessing Consensus: The Promise and Performance of Negotiated Rulemaking. *Duke Law Journal* 46: 1255–1349.

———. 1999a. The Limits of Consensus. *Environment* 41(3): 28–33.

———. 1999b. The Limits of Consensus in Environmental Regulation. Paper presented at Environmental Contracts and Regulation: Comparative Approaches in Europe and the United States. University of Pennsylvania Law School, Philadelphia, PA.

Cohen, S.A. 1979. Citizen Participation in Bureaucratic Decision Making: With Special Emphasis on Environmental Policy Making. Ph.D. dissertation. Buffalo, NY: State University of New York at Buffalo.

Cole, H., and M. Stevens. 1996. *Learning from Success: Health Agency Effort to Improve Community Involvement in Communities Affected by Hazardous Waste Sites.* July. Upper Marlboro, MD: Henry S. Cole and Associates, Inc.

Cole, L.W. 1999. The Theory and Reality of Community-Based Environmental Decision-making: The Failure of the Tanner Act and Its Implications for Environmental Justice. *Ecology Law Quarterly* 25(4): 733–756.

Cole-Misch, S., and B. Kirschner. 1995. *Evaluating Successful Strategies for Great Lakes Remedial Action Plans. A Roundtable Discussion.* Racine, WI: Wingspread.

Columbus Health Department. 1995. *Community Environmental Management Plan.* November. Columbus, OH: Columbus Health Department.

Community Advisory Group. 1994. *Final Stage I and II Report, Waukegan Harbor Remedial Action Plan.* December. Gurnee, IL: Community Advisory Group.

Cook, R.L. 1983. An Industrial Viewpoint—Virginia Toxics Roundtable. In *Seminar Proceedings: Environmental Mediation in Canada*. Ottawa, Ontario, Canada: Environmental Mediation International.

Cormick, G.W., and L.K. Patton. 1980. Environmental Mediation: Defining the Process through Experience. In *Environmental Mediation: The Search for Consensus*, edited by L.M. Lake. Boulder, CO: Westview Press: 76–97.

Cornelio, P.S., and L.G. Grimm. 1982. A Case Study of Citizen Participation in the Planning of New Transportation Facilities for the Sunbelt States. In *Compendium of Technical Papers, Institute of Transportation Engineers 52nd Annual Symposium*. Washington, DC: Institute of Transportation Engineers.

Cortese, C.F., and L. Firth. 1997. Systematically Integrating Public Participation into Planning Controversial Projects: A Case Study. *Interact* July 3(1): 6–23.

Crismon, S. 1998a. Georgia's Red-Cockaded Woodpecker Safe Harbor and Habitat Conservation Plan. In *Improving Integrated Natural Resource Planning: Habitat Conservation Plans*. National Center for Environmental Decision-Making Research website. http://ncedr.org/casestudies/hcp.html (accessed October 14, 2001).

————. 1998b. Wisconsin's Karner Blue Butterfly Habitat Conservation Plan. In *Improving Integrated Natural Resource Planning: Habitat Conservation Plans*. National Center for Environmental Decision-Making Research website. http://ncedr.org/casestudies/hcp.html (accessed October 14, 2001).

Crocker, J., M. DuPraw, J. Kunde, and W. Potapchuk. 1996. *Negotiated Approaches to Environmental Decision Making in Communities: An Exploration of Lessons Learned*. Washington, DC: National Institute for Dispute Resolution and Coalition to Improve Management in State and Local Government.

Crosby, N., J.M. Kelly, and P. Schaefer. 1986. Citizens Panels: A New Approach to Citizen Participation. *Public Administration Review* March/April: 170–178.

Cusick, L.T. 1995. Developing a Methodology and Building a Team to Ensure Objectivity in Future Land Use Planning. In *Environmental Challenges: The Next 20 Years*. Washington, DC: National Association of Environmental Professionals.

Cvetkovich, G., and T.C. Earle. 1994. The Construction of Justice: A Case Study of Public Participation in Land Management. *Journal of Social Issues* 50(Fall): 161–178.

Davis, S. 1997. Does Public Participation Really Matter in Public Lands Management? Some Evidence from a National Forest. *Southeastern Political Review* 25: 253–279.

Delli Priscoli, J.D. 1988. Conflict Resolution in Water Resources: Two 404 General Permits. *Journal of Water Resources and Planning* 114(1): 66–77.

Di Santo, D.L. 1998. Public Participation and Environmental Justice: Involving the Public at Two Superfund Sites. M.S. thesis. Tucson, AZ: The University of Arizona.

Doubleday, J., R.K. Godwin, and K. Orange. 1977. *Citizen Participation in Planning for Coastal Zone Management*. Corvallis, OR: Oregon State University, Sea Grant Program.

Dryzek, J.S. 1997. *The Politics of the Earth: Environmental Discourses*. Oxford, U.K.: Oxford University Press.

Duane, T.P. 1997. Community Participation in Ecosystem Management. *Ecology Law Quarterly* 24(4): 771–797.

Ducsik, D.W., and T.D. Austin. 1986. Open Power Plant Siting: The Pioneering (and Successful) Experience of Northern States Power. In *Public Involvement in Energy Facility Planning: The Electric Utility Experience*, edited by D.W. Ducsik. Boulder, CO: Westview Press, 273–317.

Duffield, J., and S. Depoe. 1997. Lessons from Fernald: Reversing NIMBYism through Democratic Decisionmaking. *Risk Policy Report* February 21: 31–34.

Edgar, S.L. 1990a. Case Study 2: Fitchburg Water Supply Mediation. In *Environmental Disputes: Community Involvement in Conflict Resolution,* edited by J.E. Crowfoot and J.M. Wondolleck. Washington, DC: Island Press.

———. 1990b. Case Study 7: Wisconsin Groundwater Legislation Negotiations. In *Environmental Disputes: Community Involvement in Conflict Resolution,* edited by J.E. Crowfoot and J.M. Wondolleck. Washington, DC: Island Press.

Edgett, R. 1996. Testing the Practicality of Grunig's Model of Two-Way Symmetrical Communication: A Case Study. Unpublished manuscript. Syracuse, NY: Syracuse University, Public Communications Department.

———. 1997. A Test of Q Methodology as a Tool in the Mutual Gains Approach to Communication. Unpublished manuscript. Syracuse, NY: Syracuse University, Public Communications Department.

English, M., J. Peretz, and M. Manderschied. 1999. *Smart Growth for Tennessee Towns and Counties: A Process Guide.* February. Knoxville, TN: University of Tennessee–Knoxville; Waste Management Research and Education Institute; Energy, Environment, and Resources Center.

English, M., A. Gibson, D. Feldman, and B. Tonn. 1993. *Stakeholder Involvement: Open Processes for Reaching Decisions about the Future Uses of Contaminated Sites—Final Report.* September. Knoxville, TN: University of Tennessee–Knoxville; Waste Management Research and Education Institute; Energy, Environment, and Resources Center.

English, M., M. Schweitzer, S. Schexnayder, and J. Altman. 1994. *Making a Difference: Ten Case Studies of DSM/IRP Interactive Efforts and Related Advocacy Group Activities.* Oak Ridge, TN: University of Tennessee–Knoxville; Waste Management Research and Education Institute Energy, Environment, and Resources Center.

Enterprise for the Environment. 1997. *The Environmental Protection System in Transition: Toward a More Desirable Future.* Washington, DC: Center for Strategic and International Studies.

Euler, G.M. 1996. Scenery as Policy: Public Involvement in Developing a Management Plan for the Scenic Resources of the Columbia River Gorge (National Scenic Areas, Oregon). Ph.D. dissertation. Portland, OR: Portland State University.

Finney, C., and R.E. Polk. 1995. Developing Stakeholder Understanding, Technical Capability, and Responsibility: The New Bedford Harbor Superfund Forum. *Environmental Impact Assessment Review* 15: 517–541.

Fiorino, D.J. 1990. Citizen Participation and Environmental Risk: A Survey of Institutional Mechanisms. *Science, Technology, and Human Values* 15(2): 226–243.

Fiorino, D.J., and C. Kirtz. 1985. Breaking Down Walls: Negotiated Rulemaking at EPA. *Temple Environmental Law and Technology Journal* 4: 29–40.

Fischer, F. 1993. Citizen Participation and the Democratization of Policy Expertise: From Theoretical Inquiry to Practical Cases. *Policy Sciences* 26: 165–187.

Fischoff, B. 1999. Forward. In *Social Trust and the Management of Risk,* edited by G. Cvetkovich and R. Lofstedt. London, U.K.: Earthscan Publications Ltd.

Fitzhugh, J., and D. Dozier. 1996. Finding the Common Good: Sugarbush Water Withdrawal. Forthcoming chapter for *Finding the Common Good: Case Studies in Consensus-Building and the Resolution of Natural Resource Controversies,* edited by K. Lowry and P. Adler. http://emis.com/publications.htm (accessed Nov. 20, 2001).

Flynn, S.G. 1992. The Timber/Fish/Wildlife Agreement: A Case Study of Alternative Environmental Dispute Resolution. M.S thesis. Burnaby, B.C.: Simon Fraser University.

Ford, D.S. 1986. Public Participation in the Tennessee Valley Authority's Energy Planning Process, 1983. In *Public Involvement in Energy Facility Planning: The Electric Utility Experience,* edited by D.W. Ducsik. Boulder, CO: Westview Press: 199–216.

Freeman, J. 1997. Collaborative Governance in the Administrative State. *UCLA Law Review* 45(1): 1–98.

Funk, W. 1987. When Smoke Gets in Your Eyes: Regulatory Negotiation and the Public Interest—EPA's Woodstove Standards. *Environmental Law* 18(3): 55–98.

Gauna, E. 1998. The Environmental Justice Misfit: Public Participation and the Paradigm Paradox. *Stanford Environmental Law Journal* 17(1): 3–72.

Geisler, M., P. Glover, E. Zieroth, and G. Payton. 1994. Citizen Participation in Natural Resource Management. In *Expanding Horizons of Forest Ecosystem Management: Proceedings of the Third Habitat Futures Workshop, October 1992,* edited by M.H. Huff, S.E. McDonald, H. Gucinski, L.K. Norris, and J.B. Nyberg. FSGTR-PNW-336. Portland, OR: U.S. Forest Service, Pacific Northwest Research Station, 87–100.

Georgia Times-Union. 1999. Mine Deal Costly. February 6.

Godschalk, D.R., and B. Stiffel. 1981. Making Waves: Public Participation in State Water Planning. *Journal of Applied Behavioral Science* 17(4): 597–614.

Gould, K.A. 1991a. Money, Management, and Manipulation: Environmental Mobilization in the Great Lakes Basin. Ph.D. dissertation. Evanston, IL: Northwestern University.

———. 1991b. The Sweet Smell of Money: Economic Dependency and Local Environmental Political Mobilization. *Society and Natural Resources* 4: 133–150.

Gould, K.A., A. Schnaiberg, and A. Weinberg. 1996. *Local Environmental Struggles: Citizen Activism in the Treadmill of Production.* Cambridge, U.K.: Cambridge University Press.

Graves, P.F., and W.F. LaPage. 1977. Participant Satisfaction with Public Involvement in U.S. Forest Service Recreation Policy. In *Involvement and Environment, Proceedings of the Canadian Conference on Public Participation,* edited by B. Sadler. Edmonton, Alberta, Cananda: Environment Council of Alberta.

Gray, B. 1989. *Collaborating: Finding Common Ground for Multiparty Problems.* San Francisco, CA: Jossey-Bass Publishers.

Gray, B., and T.M. Hay. 1986. Political Limits to Interorganizational Consensus and Change. *Journal of Applied Behavioral Science* 22(2): 95–112.

Gray, B., and J. Purdy. 1992. *Promised Land State Park: A Case Study of Environmental Mediation.* Case Study Series 92–1. University Park, PA: Center for Research in Conflict and Negotiation.

Gray, C., and S. Langton. 1995. *A Public Participation Bibliography.* September. Prepared for the International Association of Public Participation Practitioners. Alexandria, VA: International Association of Public Participation Practitioners.

Greenberg, M., K. Lowrie, D. Krueckeberg, H. Mayer, and D. Simon. 1997. Bombs and Butterflies: A Case Study of the Challenges of Post Cold War Environmental Planning and Management for the U.S. Nuclear Weapons Sites. *Journal of Environmental Planning and Management* 40(6): 739–750.

Greer, L.E. 1998. Why We Didn't Sign. *The Environmental Forum* March/April: 37–38.

Gregory, R. 2000. Using Stakeholder Values to Make Smarter Environmental Decisions. *Environment* 24(5): 34–44.

Gricar, B., and A.J. Baratta. 1983. Bridging the Information Gap at Three Mile Island: Radiation Monitoring by Citizens. *Journal of Applied Behavioral Science* 19(1): 35–49.

Gross, D.W. 1996. Professional Collaborative Intervention as a Technique for Social Capital Formation in Protected Landscape: A Study of a Strategy Utilized in Exmoor National Park and the Adirondack Park. Ph.D. dissertation. Ithaca, NY: Cornell University.

Gurtner-Zimmermann, A. 1994. Ecosystem Approach to Planning in the Great Lakes: A Mid-Term Review of Remedial Action Plans. Ph.D. dissertation. Toronto, Ontario, Canada: University of Toronto.

———. 1996. Analysis of Lower Green Bay and Fox River, Collingwood Harbour, Spanish Harbour, and the Metro Toronto and Region Remedial Action Plan (RAP) Process. *Environmental Management* 20(4): 449–459.

Hartig, J.H., and D.M. Dolan. 1995. Successful Application of an Ecosystem Approach—the Restoration of Collingwood Harbour. *Journal of Great Lakes Research* 21(3): 285–286.

Hartig, J.H., and N.L. Law. 1993. *Institutional Frameworks to Direct the Development and Implementation of Remedial Action Plans.* Detroit, MI: Wayne State University.

———. 1994. *Progress in Great Lakes Remedial Action Plans: Implementing the Ecosystem Approach in Great Lakes Areas of Concern.* Detroit, MI: Wayne State University.

Hartig, J.H., and M.A. Zarull (eds.). 1992. *Under RAPs: Towards Grassroots Ecological Democracy in the Great Lakes Basin.* Ann Arbor, MI: University of Michigan Press.

Hartig, J.H., R.L. Thomas, and E. Iwachewski. 1996. Lessons from Practical Application of an Ecosystem Approach in Management of the Laurentian Great Lakes. *Lakes and Reservoirs: Research and Management* 2: 137–145.

Hartley, T.W. 1998. Participant Competencies in Deliberative Discourse: Cases of Collaborative Decision-Making in the U.S. EPA Superfund Program. Paper presented at the Seventh International Symposium on Society and Natural Resource Management. May 27–31, University of Missouri–Columbia.

———. 1999. Versatility, Patience, and Non-Defensiveness: Helping Participants Cope in Collaborative Environmental Decision-Making. Unpublished manuscript on file with authors.

Haygood, L.V. 1995. Balancing Conservation and Development in Chiwaukee Prairie, Wisconsin. In *Collaborative Planning for Wetlands and Wildlife,* edited by D.R. Porter and D.A. Salvesen. Washington, DC: Island Press.

Hays, S.P. 1959. *Conservation and the Gospel of Efficiency: The Progressive Conservation Movement, 1890–1920.* Cambridge, MA: Harvard University Press.

HCEPP (Hamilton County Environmental Priorities Project). 1998a. *Issues Assessment Summary.* Hamilton County, OH: HCEPP.

———. 1998b. *Public Involvement Report.* Hamilton County, OH: HCEPP.

Hetherington, A., and L. Piotrowski. 1996. Democracy in the Woods. *Yes! A Journal of Positive Futures* Fall: 41–43.

Hoffman, S.M., and S. Matisone. 1997. *Citizen's Jury on Minnesota's Electricity Future.* Minneapolis, MN: Jefferson Center.

Holowaty, M.O., M. Reshkin, M. Mikulka, and R. Tolpa. 1992. Working toward a Remedial Action Plan for the Grand Calumet River and Indiana Harbor Ship Canal. In *Under RAPs: Toward Grassroots Ecological Democracy in the Great Lakes Basin,* edited by J.H. Hartig and M.A. Zarull. Ann Arbor, MI: University of Michigan Press, 121–138.

Houghton, A., and L. Siegel. 1997. *The Fort Ord Restoration Advisory Board Interim Report and Recommendations.* July. San Francisco, CA: CAREER/PRO.

Howell, R.E., M. Olsen, and D. Olsen. 1987. *Designing a Citizen Involvement Program: A Guidebook for Involving Citizens in the Resolution of Environmental Issues.* Corvallis, OR: Oregon State University, Western Rural Development Center.

IJC (International Joint Commission). 1995. *Evaluating Successful Strategies for Great Lakes Remedial Action Plans.* Detroit, MI: IJC.

———. 1996. *Great Lakes Water Quality Board Position Statement on the Future of Great Lakes Remedial Action Plans.* Detroit, MI: IJC.

———. 1998. *Beacons of Light: Special Report on Successful Strategies toward Restoration in Areas of Concern under the Great Lakes Water Quality Agreement.* Detroit, MI: IJC.

Innes, J.E. 1998. Information in Communicative Planning. *Amercan Planning Association Journal* 64(1): 52–63.

Institute for Local Government Administration and Rural Development. 1995. *Environmental Priorities: Athens County.* August. Athens, OH: Ohio University.

International Symposium on Technology and Society. 1996. Technical Expertise and Public Decisions. *Proceedings of 1996 International Symposium on Technology and Society.* Princeton, NJ: Princeton University.

Jaffray, B. 1981. *Public Involvement in Environmental Decisionmaking: An Annotated Bibliography.* August. Toronto, Ontario, Canada: Council of Planning Librarians Bibliographies.

Jefferson Center. 1985. *Final Report of the Citizens Panel on Agriculture and Water Quality.* Minneapolis, MN: Jefferson Center.

———. 1995. *Report on Traffic Congestion Pricing.* Minneapolis, MN: Jefferson Center.

———. 1996a. *Citizens Jury on Comparing Environmental Risks: Final Report.* Minneapolis, MN: Jefferson Center.

———. 1996b. *Citizens Jury on Hog Farming: Final Report.* Minneapolis, MN: Jefferson Center.

———. 1997. *Citizens Jury on Dakota County's Comprehensive Land Use.* Minneapolis, MN: Jefferson Center.

John, D., and M. Mlay. 1997. *Reaching Environmental Solutions at the Community Level: Lessons from the Upper Clark Fork River.* February 4. Washington, DC: National Academy of Public Administration. Unpublished draft.

Johnson, K. 1998. The Henry's Fork Watershed Council: Community-Based Participation in Regional Environmental Management. In *Sustainable Community Development: Studies in Economic, Environmental, and Cultural Revitalization,* edited by M.D. Hoff. Boca Raton, FL: Lewis Publishers.

Johnson, K., M.D.B. Delson, M. DuPraw, and J. Crocker. 1996. Collaborating to Build Sustainable Communities. Working paper. April. Washington, DC: Program for Community Problem-Solving.

Johnson, P. 1993. How I Turned a Critical Public into Useful Consultants. *Harvard Business Review* (January–February): 56–66.

Johnston, A.M. 1999. Public Participation in the Politics of Environmental Discourse: The Anitech Image Corporation Pollution Controversy. Ph.D. dissertation. Ithaca, NY: Cornell University.

Kaemmerer, D., A. O'Brien, T. Sheffy, and S. Skavroneck. 1992. The Quest for Clean Water: The Milwaukee Estuary Remedial Action Plan. In *Under RAPs: Toward Grassroots Ecological Democracy in the Great Lakes Basin,* edited by J.H. Hartig and M.A. Zarull. Ann Arbor, MI: University of Michigan Press, 139–160.

Kalikow, B. 1984. Environmental Risk: Power to the People. *Technology Review* 87: 54–61.

Kaminstein, D.S. 1996. Persuasions in a Toxic Community: Rhetorical Aspects of Public Meetings. *Human Organization* 55(4): 458–464.

Kathlene, L., and J.A. Martin. 1991. Enhancing Citizen Participation: Panel Designs, Perspectives, and Policy Formation. *Journal of Policy Analysis and Management* 10(1): 46–63.

Kauffman, S.E. 1992. The Causes of Conflict and Methods of Resolution in a Citizen Participation Program: A Case Study of the Lipari Landfill Superfund Site. Ph.D. dissertation. Bryn Mawr, PA: Bryn Mawr College.

Kaufman, B.L. 1998. Environmentalists Pick Top 3. *Cincinnati Enquirer,* June 7, 1998.

Kellogg, W. (ed.). 1993a. *Connecting Channel Areas of Concern: Strategies for Implementing Remedial Action Plans.* Monograph no. 5. Buffalo, NY: State University of New York at Buffalo.

———. 1993b. Ecology and Community in the Great Lakes Basin: The Role of Stakeholders and Advisory Committees in the Environmental Planning Process. Ph.D. dissertation. Ithaca, NY: Cornell University.

Kenney, D.S. 1997. *Resource Management at the Watershed Level: An Assessment of the Changing Federal Role in the Emerging Era of Community-Based Watershed Management.* Report to the Western Water Policy Review Advisory Committee. Boulder, CO: University of Colorado School of Law, Natural Resources Law Center.

Kerwin, C. 1999. *Rulemaking: How Government Agencies Write Law and Make Policy.* Washington, DC: Congressional Quarterly, Inc.

Kraft, M.E. 1995. Citizens, Experts, and Democratic Dialogue: Determining Acceptable Risks. Paper presented at the Annual Meeting of the American Political Science Association. Aug. 30, Chicago, IL.

Kraft, M.E., and B.N. Johnson. 1998. Clean Water and the Promise of Collaborative Decisionmaking: The Case of the Fox-Wolf Basin in Wisconsin. In *Toward Sustainable Communities: Transition and Transformations in Environmental Policy,* edited by D.A. Mazmanian and M.E. Kraft. Cambridge, MA: MIT Press.

Krimsky, S., and A. Plough. 1988. Plant Closure: The ASARCO/Tacoma Copper Smelter. In *Environmental Hazards: Communicating Risks as a Social Process,* edited by A. Plough and S. Krimsky. New York, NY: Auburn House: 180–237.

Laird, F.N. 1993. Participatory Analysis, Democracy, and Technological Decision Making. *Science, Technology, and Human Values* 18(3): 341–361.

Lake, L. 1980a. Mediating the West Side Highway Dispute in New York City. In *Environmental Mediation: The Search for Consensus,* edited by L.M. Lake. Boulder, CO: Westview Press, 205–233.

———. 1980b. Participatory Evaluations of Energy Options for California: A Case Study in Conflict Avoidance. In *Environmental Mediation: The Search for Consensus,* edited by L.M. Lake. Boulder, CO: Westview Press, 147–149.

Lampe, D., and M. Kaplan. 1999. Resolving Land-Use Conflicts through Mediation: Challenges and Opportunities. Working paper. Cambridge, MA: Lincoln Institute of Land Policy.

Landre, B.K. 1991. Evaluating Public Involvement in Remedial Action Planning for Great Lakes Areas of Concern. M.S. thesis. Ithaca, NY: Cornell University.

Landre, B.K., and B.A. Knuth. 1990. *Public Participation in Great Lakes Remedial Action Planning.* Ithaca, NY: New York Sea Grant Extension.

————. 1993. Success of Citizen Advisory Committees in Consensus-Based Water Resources Planning in the Great Lakes Basin. *Society and Natural Resources* 6(3): 229–257.

Langton, S. 1996. *An Organizational Assessment of the U.S. Army Corps of Engineers in Regard to Public Involvement Practices and Challenges*. Reston, VA: U.S. Army Corps of Engineers, Institute for Water Resources.

Larrson, R.A. 1993. Case Survey Methodology: Quantitative Analysis of Patterns across Case Studies. *Academy of Management Journal* 36(6): 1515–1546.

Lauber, T.B., and B. Knuth. 1996. *Fairness and Moose Management Decision-Making: The Citizens' Perspective*. Ithaca, NY: Cornell University, Department of Natural Resources, Human Dimensions Research Unit.

Leach, W.D., N.W. Pelkey, and P.A. Sabatier. 2000. Conceptualizing and Measuring Success in Collaborative Watershed Partnerships. Paper presented at the 2000 Annual Meeting of the American Political Science Association. Aug. 31–Sept. 3, Washington, DC.

Lefkoff, M.S. 1995. *Use of a Facilitated Task Force to Develop a General Permit in Colorado*. Reston, VA: U.S. Army Corps of Engineers, Institute for Water Resources.

Letterhos, J.A. 1992. Dredging up the Past: The Challenge of the Ashtabula River Remedial Action Plan. In *Under RAPs: Toward Grassroots Ecological Democracy in the Great Lakes Basin,* edited by J.H. Hartig and M.A. Zarull. Ann Arbor, MI: University of Michigan Press, 121–138.

Linscott, N. 1994. Case History: The Southern Water Supply Project. Working paper 94–52. February. Boulder, CO: University of Colorado Conflict Research Consortium, 1–10.

Loker, C.A., and D. Decker. 1994. *The Colorado Black Bear Hunting Controversy: A Case Study of Human Dimensions in Contemporary Wildlife Management*. Ithaca, NY: Cornell University, Department of Natural Resources, Human Dimensions Research Unit.

Lucas, W.A. 1974. *The Case Survey Method: Aggregating Case Experience*. RAND report no. R-1515-RC. October. Santa Monica, CA: RAND.

Lynn, F. 1987. Citizen Involvement in Hazardous Waste Sites: Two North Carolina Success Stories. *Environmental Impact Assessment Review* 18: 347–361.

Lynn, F.M., and G.J. Busenberg. 1995. Citizen Advisory Committees and Environmental Policy: What We Know, What's Left to Discover. *Risk Analysis* 15(2): 147–162.

MacKenzie, S.H. 1993. Ecosystem Management in the Great Lakes: Some Observations from Three RAP Sites. *Journal of Great Lakes Research* 19(1): 136–144.

————. 1996. *Integrated Resource Planning and Management: The Ecosystem Approach in the Great Lakes Basin*. Washington, DC: Island Press.

Mangerich, M.K., and L.S. Luton. 1995. The Inland Northwest Field Burning Summit: A Case Study. In *Mediating Environmental Conflicts,* edited by J.W. Blackburn. Westport, CT: Quorum, chapter 17, 247–265.

Marks, J. 1999. Consent Building Gets Wastewater Project Back on Track. *Participation Quarterly* (First Quarter).

Marriott, D.R. n.d. Integrating Science and Public Values in Environmental Planning: A Case Study of the Lower Columbia River Estuary Program. Unpublished manuscript. Portland, OR: Lower Columbia River Estuary Program.

Mazmanian, D.A., and J. Nienaber. 1979. *Can Organizations Change? Environmental Protection, Citizen Participation, and the Army Corps of Engineers*. Washington, DC: Brookings Institution.

McComas, K.A., and C.W. Scherer. 1998. Reassessing Public Meetings as Participation in Risk Management Decisions. *Risk: Health, Safety, and Environment* 9(4): 361–378.

McConnon, D. 1986. Public Participation in Routing Transmission Lines: A Program Born of Adversity. In *Public Involvement in Energy Facility Planning: The Electric Utility Experience*, edited by D.W. Ducsik. Boulder, CO: Westview Press.

McGlennon, J. 1983. The Process of Mediation and the Patuxent River Cleanup Agreement. In *Seminar Proceedings: Environmental Mediation in Canada*. Ottawa, Ontario, Canada: Environmental Mediation International.

McMullin, S.L., and L.A. Nielsen. 1991. Resolution of Natural Resource Allocation Conflicts through Effective Public Involvement. *Policy Studies Journal* 19(3–4): 553–564.

McNelly, P.W. 1982. Citizen Participation in Wastewater Sludge Management Planning: A Study of Apathy and Protest. M.S. thesis. Fullerton, CA: California State University–Fullerton.

Menton, D. (ed.). 1996. *Siting by Choice: Waste Facilities, NIMBY, and Volunteer Communities*. Washington, DC: Georgetown University Press.

Mernitz, S. 1980. *Mediation of Environmental Disputes: A Source Book*. New York, NY: Praeger Publishers.

Merrick, J. 1998a. The Louisiana Black Bear Conservation Plan. In *Improving Integrated Natural Resource Planning: Habitat Conservation Plans*. National Center for Environmental Decision-Making Research website. http://www.ncedr.org/casestudies/hcp.html (accessed October 14, 2001).

———. 1998b. San Diego Multiple Species Conservation Plan. In *Improving Integrated Natural Resource Planning: Habitat Conservation Plans*. National Center for Environmental Decision-Making Research website. http://www.ncedr.org/casestudies/hcp.html (accessed October 14, 2001).

———. 1998c. Tulare County Habitat Conservation Plan. In *Improving Integrated Natural Resource Planning: Habitat Conservation Plans*. National Center for Environmental Decision-Making Research website. http://www.ncedr.org/casestudies/hcp.html (accessed October 14, 2001).

Michaud, G.R. 1998. *Economic Benefits from Sediment Cleanup: The Case in Waukegan Harbor*. Chicago, IL: Illinois Environmental Protection Agency.

Michigan Area of Concern News. 1992. *Michigan Area of Concern News*, Fall issue.

———. 1997. *Michigan Area of Concern News*, Spring and Fall issues.

Michigan Department of Natural Resources. 1994. *Rouge River Remedial Action Plan Update*. Detroit, MI: Michigan Department of Natural Resources.

Michigan RRAP (Relative Risk Analysis Project). 1992. *Michigan's Environment and Relative Risk*. Detroit, MI: Michigan Department of Environmental Quality.

Miller, M. 1998a. North Carolina Sandhills Safe Harbor Plan. In *Improving Integrated Natural Resource Planning: Habitat Conservation Plans*. National Center for Environmental Decision-Making Research website. http://ncedr.org/casestudies/hcp.html (accessed October 14, 2001).

———. 1998b. Plum Creek Habitat Conservation Plan. In *Improving Integrated Natural Resource Planning: Habitat Conservation Plans*. National Center for Environmental Decision-Making Research website. http://ncedr.org/casestudies/hcp.html (accessed October 14, 2001).

Miller, R.C., and T.R. Griffith. 1994. Gaining Credibility on a Controversial High-Speed Rail Project through the NEPA Process: A Case Study. *Journal of Environmental Permitting* 4(1): 59–75.

Minard, R., K. Jones, and C. Paterson. 1993. *State Comparative Risk Projects: A Force for Change*. Montpelier, VT: Green Mountain Institute.

Miranda, M.L., J. Miller, and T.L. Jacobs. 1996. Informing Policy Makers and the Public in Landfill Siting Processes. In *Technical Expertise and Public Decisions: Proceedings of 1996 International Symposium on Technology and Society.* Princeton, NJ: Princeton University.

Mogen, R. 1986. Citizen Participation in Decision-Making at Portland General Electric. In *Public Involvement in Energy Facility Planning: The Electric Utility Experience,* edited by D.W. Ducsik. Boulder, CO: Westview Press, 177–197.

Moore, C.W. 1991. *Corps of Engineers Uses Mediation to Settle Hydropower Dispute.* Reston, VA: U.S. Army Corps of Engineers, Institute for Water Resources.

Moore, L. 1997. From Public Involvement to Mediation: A Cautionary Tale. *Interact* 3(1): 57–70.

Moote, M.A., M.P. McClaran, and D.K. Chickering. 1997. Theory in Practice: Applying Participatory Democracy Theory to Public Land Planning. *Environmental Management* 21(6): 877–889.

NAEP (National Association of Environmental Professionals). 1998. *Overcoming Barriers to Environmental Improvement: Conference Proceedings.* Bowie, MD: NAEP.

———. 1999. *Environment in the 21st Century: Conference Proceedings.* Bowie, MD: NAEP.

NAPA (National Academy of Public Administration). 1997. *Excellence, Leadership, and the Intel Corporation: A Study of EPA's Project XL. In Resolving the Paradox of Environmental Protection.* Washington, DC: National Academy of Public Administration.

NCCR (Northeast Center for Comparative Risk). 1995. A Provisional Ranking for Northeast Ohio. *The Comparative Risk Bulletin* 5(3 & 4).

———. 1996a. Life after Ranking: Columbus Presses On. *The Comparative Risk Bulletin* 6(1 & 2).

———. 1996b. Columbus Means Business. *The Comparative Risk Bulletin* 6(3 & 4).

NEJAC (National Environmental Justice Advisory Committee). 1996. *The Model Plan for Public Participation.* Washington, DC: U.S. Environmental Protection Agency.

Nelson, K.C. 1990a. Case Study 3: Common Ground Consensus Project. In *Environmental Disputes: Community Involvement in Conflict Resolution,* edited by J.E. Crowfoot and J.M. Wondolleck. Washington, DC: Island Press.

———. 1990b. Case Study 5: Sand Lakes Quiet Area Issue-Based Negotiation. In *Environmental Disputes: Community Involvement in Conflict Resolution,* edited by J.E. Crowfoot and J.M. Wondolleck. Washington, DC: Island Press.

———. 1990c. Case Study 6: Pig's Eye Attempted Mediation. In *Environmental Disputes: Community Involvement in Conflict Resolution,* edited by J.E. Crowfoot and J.M. Wondolleck. Washington, DC: Island Press.

Neuman, J.C. 1996. Run, River, Run: Mediation of a Water-Rights Dispute Keeps Fish and Farmers Happy—for a Time. *University of Colorado Law Review* 67: 259–340.

NNOTF (New National Opportunities Task Force). 1997. *Lessons Learned from Collaborative Approaches.* Washington, DC: President's Council on Sustainable Development.

Ohio Comparative Risk Project. 1995a. *A Guide to Comparative Risk.* Columbus, OH: Ohio Environmental Protection Agency.

———. 1995b. *Ohio State of the Environment Report.* Columbus, OH: Ohio Environmental Protection Agency.

———. 1997. *Recommendations to Reduce Environmental Risk in Ohio.* Columbus, OH: Ohio Environmental Protection Agency.

Ohio EPA (Environmental Protection Agency). 1991. *Ashtabula River Remedial Action Plan, Stage 1 Report, Background Investigation.* Columbus, OH: Ohio EPA.

One Day Conference on Risk. 1997. Abstracts of meeting. June 13, City University, London, U.K. Society for Risk Analysis website. http://www.riskworld.com/Abstract/ab7me002.htm.

Onibokun, A., and M. Curry. 1976. An Ideology of Citizen Participation: The Metropolitan Seattle Transit Case Study. *Public Administration Review* May/June: 269–277.

Opperman, J. 1998a. Maine's Atlantic Salmon Conservation Plan. In *Improving Integrated Natural Resource Planning: Habitat Conservation Plans*. National Center for Environmental Decision-Making Research website. http://ncedr.org/casestudies/hcp.html (accessed October 14, 2001).

———. 1998b. The Pleasant Valley Habitat Conservation Planning Effort. In *Improving Integrated Natural Resource Planning: Habitat Conservation Plans*. National Center for Environmental Decision-Making Research website. http://ncedr.org/casestudies/hcp.html (accessed October 14, 2001).

Orenstein, S.G. 1998. *Evaluation of Project XL Stakeholder Processes*. Washington, DC: RESOLVE, Inc.

Osterman, D., F. Steiner, T. Hicks, R. Ledgerwood, and K. Gray. 1989. Coordinated Resource Management and Planning: The Case of the Missouri Flat Creek Watershed. *Journal of Soil and Water Conservation* 44(5): 403–406.

Ozawa, C.P. 1991a. *Recasting Science: Consensual Procedures in Public Policy Making*. Boulder, CO: Westview Press.

———. 1991b. Transformative Mediation Techniques: Improving Public Participation in Environmental Decision Making. Working paper 91–7, Cambridge, MA: Harvard Law School, Program on Negotiation.

Paterson, C.J., and R.N. Andrews. 1995. Procedural and Substantive Fairness in Risk Decisions: Comparative Risk Assessment Procedures. *Policy Studies Journal* 23(1): 85–95.

Patterson, K., P.K. Smith, and D. Martin. 1998. A Balancing Act: How Opposing Positions Developed Meaningful Public Forums. Paper presented at the Third Topical Meeting: DOE Spent Nuclear and Fissile Materials Management, American Nuclear Society. Sept. 9–11, Charleston, SC.

Peelle, E., M. Schweitzer, J. Munro, S. Carnes, and A. Wolfe. 1996. Factors Favorable to Public Participation Success. In *Practical Environmental Directions: A Changing Agenda, National Association of Environmental Professionals 21st Annual Conference Proceedings*. Washington, DC: National Association of Environmental Professionals.

Perhac, R.M. 1996. Defining Risk: Normative Considerations. *Human and Ecological Risk Assessment* 2(2): 381–392.

Perritt, H.J. 1986. Negotiated Rulemaking before Federal Agencies: Evaluation of Recommendations by the Administrative Conference of the United States. *Georgetown Law Journal* 74: 1625–1692.

———. 1995. Use of Negotiated Rulemaking to Develop a Proposed OSHA Health Standard for MDA. In *Negotiated Rulemaking Sourcebook,* edited by D.M. Pritzker and D.S. Dalton. Washington, DC: Administrative Conference of the United States.

Pindyck, R.S., and D.L. Rubinfeld. 1991. *Economic Models and Economic Forecasts*. New York, NY: McGraw–Hill.

Platt, R.H. 1995. The 2020 Water Supply Study for Metropolitan Boston: The Demise of Diversion. *Journal of the American Planning Association* 61(2): 185–199.

Plumlee, J.P., J.D. Starling, and K.W. Kramer. 1985. Citizen Participation in Water Quality Planning: A Case Study of Perceived Failure. *Administration and Society* 16(4): 455–473.

Powell, J.D. 1988. A Hazardous Waste Site: The Case of Nyanza. In *Environmental Hazards: Communicating Risks as a Social Process,* edited by A. Plough and S. Krimsky. New York, NY: Auburn House, 239–297.

PRC (Pew Research Center). 1998. *Deconstructing Trust: How Americans View Government.* Washington, DC: Pew Research Center.

Press, D. 1994. *Democratic Dilemmas in the Age of Ecology: Trees and Toxics in the American West.* Durham, NC: Duke University Press.

Priorities '95 Steering Committee. 1995. *Priorities '95 Final Report and Strategic Recommendations.* Columbus, OH: Priorities '95 Steering Committee.

Pritzker, D.M., and D.S. Dalton (eds.). 1995. *Negotiated Rulemaking Sourcebook.* Washington, DC: Administrative Conference of the United States.

Purdy, J.M., and B. Gray. 1994. Government Agencies as Mediators in Public Policy Conflicts. *International Journal of Conflict Management* 5(2): 158–180.

Putnam, R.D. 1995. Bowling Alone: America's Declining Social Capital. *Journal of Democracy* 6(1): 65–78.

Rabe, B.G. 1994. *Beyond NIMBY: Hazardous Waste Siting in Canada and the United States.* Washington, DC: The Brookings Institution.

Raffensperger, C. 1998. Guess Who's Coming for Dinner: The Scientist and the Public Making Good Environmental Decisions. *Human Ecology Review* 5(1): 37–41.

Reich, R.B. 1985. Public Administration and Public Deliberation: An Interpretive Essay. *Yale Law Journal* 94(7): 1617–1641.

Renn, O., T. Webler, and P. Wiedemann. 1995. *Fairness and Competence in Citizen Participation: Evaluating Models for Environmental Discourse.* Dordrecht, the Netherlands: Kluwer Academic Publishers.

RESOLVE. 1999. Collaborative Process to Address Controversy about DuPont Proposal for Titanium Mine Adjacent to Okefenokee National Wildlife Refuge. *ResolveFacts Project Description.* Washington, DC: RESOLVE, Inc.

Risk Assessment and Policy Association. 1999. Abstracts for the Second Biennial International Meeting. March 25–26, 1999, Alexandria, VA.

Rogers, E., L. Murakami, and L. Hanson. 1995. *How Citizen Advisory Boards Provide Input into Major Waste Policy Decisions.* Westminster, CO: Rocky Flats Citizens Advisory Board.

Rosener, J. 1983. *User-Oriented Evaluation: A New Way to View Citizen Participation. Public Involvement and Social Impact Assessment.* Boulder, CO: Westview Press, 45–60.

Ross, P., L. Burnett, and C. Davis. 1992. Remediating Contamination in the Waukegan, Illinois, Area of Concern. In *Under RAPs: Toward Grassroots Ecological Democracy in the Great Lakes Basin,* edited by J.H. Hartig and M.A. Zarull. Ann Arbor, MI: University of Michigan Press, 235–250.

Ruckelshaus, W.D. 1996. *Trust in Government: A Prescription for Restoration.* Washington, DC: National Academy of Public Administration.

———. 1998. Stepping Stones. *The Environmental Forum* March/April: 30–36.

Salvesen, D. 1995. Collaborative Planning for Development in Bolsa Chica, California's Wetlands. In *Collaborative Planning for Wetlands and Wildlife: Issues and Examples,* edited by D. Porter and D. Salvesen. Washington, DC: Island Press, 257–273.

Schattle, H. 1998. Challenges of Public Deliberation: A Case Study of Citizen Participation in Environmental Policy. M.A. thesis. Boston, MA: Boston College.

Scher, E. 1997. Consensual Approaches to Environmental Decision-Making: The Case of the Massachusetts Military Reservation. *Environment Reporter* 28(26): 1309–1315.

Schmiechen, P., L. Kolze, and J. Melander. 1997. *Risk-Based Environmental Priorities Project: Final Report.* St. Paul, MN: Minnesota Pollution Control Agency, Environmental Planning Unit.

Schneider, H.G. 1994. Case Study: Lake Catamount. Working paper 94–53. Boulder, CO: University of Colorado, Conflict Research Consortium.

Schneider, M., P. Teske, and M. Marschall. 1997. Institutional Arrangements and the Creation of Social Capital: The Effects of Public School Choice. *American Political Science Review* 91(1): 82–93.

Schneider, S.M., and R. Beckingham. 2000. *Changing the Nature of Online Conversation: An Evaluation of RealityCheck.com.* February. http://www.weblab.org/home.html (accessed Nov. 19, 2001).

Schrameck, R., M. Fields, and M. Synk. 1992. Restoring the Rouge. In *Under RAPs: Toward Grassroots Ecological Democracy in the Great Lakes Basin,* edited by J.H. Hartig and M.A. Zarull. Ann Arbor, MI: University of Michigan Press, 73–91.

Scott, E. 1988. *Managing Environmental Risks: The Case of Asarco.* Cambridge, MA: Harvard University, Kennedy School of Government, Case Program.

SEAB (Secretary of Energy Advisory Board). 1993. *Earning Public Trust and Confidence: Requisites for Managing Radioactive Wastes.* Washington, DC: U.S. Department of Energy, Task Force on Radioactive Waste Management.

Seltzer, E.P. 1983. Citizen Participation in Environmental Planning: Context and Consequence. Ph.D. dissertation. Philadelphia; PA: University of Pennsylvania.

Sewell, W.R., D. Phillips, and S.D. Phillips. 1979. Models for the Evaluation of Public Participation Programs. *Natural Resources Journal* 19: 337–358.

Shepherd, A., and C. Bowler. 1997. Beyond the Requirements: Improving Public Participation in EIA. *Journal of Environmental Planning and Management* 40(6): 725–738.

Sherlock, P. 1996. Idaho Learns to Share Two Rivers. *High Country News* 28(9) [online]. http://www.hcn.org/1996/May13/dir/Feature_Idaho_learn.html (accessed July 15, 1999).

Siemer, W.F., and D.J. Decker. 1990. *An Evaluation of Public Meetings Held by the DEC Bureau of Wildlife (October 1989).* Ithaca, NY: Cornell University, Department of Natural Resources, Human Dimensions Research Unit.

Simon, D. 1999. *The St. Louis FUSRAP Sites—A Preliminary Case Study.* Report 26. New Brunswick, NJ: Consortium for Risk Evaluation with Stakeholder Participation.

Sirianni, C. 1999. *The Tacoma Smelter and EPA.* Tacoma, WA: Civic Practices Network.

Slovic, P. 1992. Perceptions of Risk: Reflections on the Psychometric Paradigm. In *Social Theories of Risk,* edited by S. Krimsky and D. Golding. Westport, CT: Praeger.

———. 1993. Perceived Risk, Trust, and Democracy. *Risk Analysis* 13(6): 675–682.

Society for Risk Analysis. 1986, 1998, 1989. Abstracts of annual meetings. Washington, DC: Society for Risk Analysis.

———. 1994–1998. Abstracts of annual meetings. Society for Risk Analysis website. http://www.riskworld.com/Abstract/AB5ME001.HTM.

SPAC (Stakeholder Public Advisory Council). 1997. *Overcoming Obstacles to Sediment Remediation in the Great Lakes Basin.* Detroit, MI: Great Lakes Water Quality Board, International Joint Commission.

Stancik, M.M. 1995. Combining Tools and Processes to Facilitate Coastal Environmental Decisions Which Reflect Well-Informed Societal Preferences. Ph.D. dissertation. Cambridge, MA: Massachusetts Institute of Technology.

Stata. 1997. *Stata Reference Manual (Release 5, Volume 3, P–Z)*. College Station, TX: Stata Press.

Steelman, T. 1996. Public Participation in National Forest Management: A Case Study of the Monongahela National Forest, West Virginia. In *Technical Expertise and Public Decisions. Proceedings of 1996 International Symposium on Technology and Society*. Princeton, NJ: Princeton University.

Steelman, T.A., and W. Ascher. 1997. Public Involvement Methods in Natural Resource Policy Making: Advantages, Disadvantages, and Trade-Offs. *Policy Sciences* 30: 71–90.

Steelman, T.A., and J. Carmin. 1998. Common Property, Collective Interests, and Community Opposition to Locally Unwanted Land Uses. *Society and Natural Resources* 11(5): 485–504.

Steinzor, R., and S. Strauss. 1987. Building a Consensus: Agencies Stressing Reg-Neg Approach. *Legal Times* Aug. 3: 16–21.

Stephan, M. 1998. Grassroots Activism at Superfund Sites: The Case of the Southern Maryland Wood Treating Site. Paper presented at Annual Meeting of the Western Political Science Association. March 19–20, Los Angeles, CA.

Stern, P.C., and H.V. Fineberg (eds.). 1996. *Understanding Risk: Informing Decisions in a Democratic Society*. Washington, DC: National Academy Press.

Stewart, R.B. 1975. The Reformation of American Administrative Law. *Harvard Law Review* 88(8): 1671–1813.

Stewart, T.R., R.L. Dennis, and D.W. Ely. 1984. Citizen Participation and Judgment in Policy Analysis: A Case Study of Urban Air Quality Policy. *Policy Sciences* 17(1): 67–87.

Stokes, M.E., C. Davis, and G. Koch. 1995. *Categorical Data Analysis Using the SAS System*. Cary, NC: SAS Institute, Inc.

Stout, R.J., and B.A. Knuth. 1994. *Evaluation of a Citizen Task Force Approach to Resolve Suburban Deer Management Issues*. Ithaca, NY: Cornell University, Department of Natural Resources, Human Dimensions Research Unit.

———. 1995. *Effects of a Suburban Deer Management Communication Program, with Emphasis on Attitudes and Opinions of Suburban Residents*. Ithaca, NY: Cornell University, Department of Natural Resources, Human Dimensions Research Unit.

Sullivan, G.L. 1997. Partnerships in Practice: The Fine Line between Success and Failure. In *Transactions of the Sixty-Second North American Wildlife and Natural Resources Conference: Finding Common Ground in Uncommon Times*, edited by K.G. Wadsworth. Washington, DC: Wildlife Management Institute.

Susskind, L., and J. Cruikshank. 1987. *Breaking the Impasse: Consensual Approaches to Resolving Public Disputes*. New York, NY: Basic Books.

Susskind, L., and G. McMahon. 1985. The Theory and Practice of Negotiated Rulemaking. *Yale Journal on Regulation* 3: 133–165.

Susskind, L., and L. Van Dam. 1986. Squaring Off at the Table, Not in the Courts: All Too Often Government Regulations End Up as the Subject of Lawsuits. *Technology Review* 89: 36–44.

Susskind, L.E., S.L. Podziba, and E. Babbitt. 1989. *Goodyear Tire and Rubber Company*. Alternative Dispute Resolution Series, Case study 89-ADR-CS-5. Washington, DC: U.S. Army Corps of Engineers, Institute for Water Resources.

Susskind, L., O. Amundsen, M. Matsuura, M. Kaplan, and D. Lampe. 1999. *Using Assisted Negotiation to Settle Land Use Disputes: A Guidebook for Public Officials*. Cambridge, MA: Lincoln Institute for Land Policy.

Swearingen, S.L. 1998. Democratic Debates in Land Use Planning and Decision-Making: A Case Study of Public Participation in Wisconsin. Ph.D. dissertation. Madison, WI: University of Wisconsin–Madison.

Szaz, A., and M. Meuser. 1995. *Stakeholder Participation in the Toxic Cleanup of Military Facilities and Its Relationship to the Prospects for Economic Reuse: The Case of Fort Ord, California.* Carmel, CA: Fort Ord Toxics Project.

Tableman, M.A. 1990. Case Study 1: San Juan National Forest Mediation. In *Environmental Disputes: Community Involvement in Conflict Resolution,* edited by J.E. Crowfoot and J.M. Wondolleck. Washington, DC: Island Press.

Talbot, A.R. 1983. *Settling Things: Six Case Studies in Environmental Mediation.* Washington, DC: The Conservation Foundation.

Thomas, C.W. 2001. Habitat Conservation Planning: Certainly Empowered, Somewhat Deliberative, Questionably Democratic. *Politics and Society* 29(1): 105–130.

Thomas, J.C. 1993. Public Involvement and Governmental Effectiveness: A Decision-Making Model for Public Managers. *Administration and Society* 24(4): 444–469.

University of Cincinnati. 1998. Risk Bibliography. In *Environmental Communications Bibliographies.* University of Cincinnati, Center for Environmental Communication Studies website. http://www.uc.edu/cecs/cecs.html (accessed Nov. 19, 2001).

U.S. DOE (Department of Energy). 1997. *Site Specific Advisory Board Initiative 1997 Evaluation Survey Results.* Washington, DC: U.S. Department of Energy, Office of Environmental Management.

U.S. EPA (Environmental Protection Agency). 1996. *Community Advisory Groups: Partners in Decisions at Hazardous Waste Sites.* Washington, DC: U.S. EPA, Office of Solid Waste and Emergency Response, Community Involvement and Outreach Center.

———. 2000. *Baltimore Community Environmental Protection Air Committee Technical Report.* EPA report 744-R-00–005. April. Washington, DC: U.S. EPA, Office of Pollution Prevention and Toxics.

U.S. EPA (Environmental Protection Agency) SAB (Science Advisory Board). 2001. *Improved Science-Based Environmental Stakeholder Processes: A Commentary by the EPA Science Advisory Board.* EPA-SAB-EC-COM-01-006. Washington, DC: U.S. EPA SAB.

Valdez, J.K. 1993. The Emergence and Dynamics of Citizen Participation in Wyoming Energy Conflicts in the 1970s. Ph.D. dissertation. Madison, WI: University of Wisconsin–Madison.

WCED (Western Center for Environmental Decision-Making). 1997. *Public Involvement in Comparative Risk Projects: Principles and Best Practices.* Boulder, CO: Western Center for Environmental Decision-Making.

Weber, E.P. 1998. Successful Collaboration: Negotiating Effective Regulation. *Environment* 40(9): 10–15.

———. 2000. A New Vanguard for the Environment: Grass-Roots Ecosystem Management as a New Environmental Movement. *Society and Natural Resources* 13(3): 237–259.

Weber, E.P., and A.M. Khademian. 1997. From Agitation to Collaboration: Clearing the Air through Negotiation. *Public Administration Review* 57(5): 396–410.

Webler, T.N. 1992. Modeling Public Participation as Discourse: An Application of Habermas' Theory of Communicative Action. Ph.D. dissertation. Worcester, MA: Clark University.

Wernstedt, K., and R. Hersh. 1997. Land Use and Remedy Selection: Experience from the Field—the Fort Ord Site. Discussion paper 97–28. Washington, DC: Resources for the Future.

Whitlatch, E.E., and J.A. Aldrich. 1980. *Energy Facility Siting Procedures, Criteria, and Public Participation in the Ohio River Basin Energy Study Region.* Washington, DC: U.S. Environmental Protection Agency, Office of Research and Development.

Williams, B.A., and A.R. Matheny. 1995. *Democracy, Dialogue, and Environmental Disputes: The Contested Languages of Social Regulation.* New Haven, CT: Yale University Press.

Wilson, J. 1993. *Improving Toxic Materials Conflicts through Improved Public Participation.* Boulder, CO: University of Colorado, Conflict Research Consortium.

Yin, R.K., and K.A. Heald. 1975. Using the Case Survey Method to Analyze Policy Studies. *Administrative Science Quarterly* 20: 371–381.

Yosie, T.F., and T.D. Herbst. 1998. *Using Stakeholder Processes in Environmental Decisionmaking: An Evaluation of Lessons Learned, Key Issues, and Future Challenges.* Washington, DC: Ruder Finn Washington.

Index

About the Authors

Thomas C. Beierle is a fellow at Resources for the Future. His research focuses on the role of the public in environmental decisionmaking and the evaluation of public participation processes. He has published articles in the *Journal of Policy Analysis and Management, Policy Studies Review, Environment and Planning, Society and Natural Resources, Environment,* and *The Environmental Law Reporter.*

Jerry Cayford is a research associate at Resources for the Future. A philosopher by training, he wrote his dissertation on the philosophy of Ludwig Wittgenstein. His public policy work concerns the public's perceptions of risk and public participation in environmental decisions. Most recently, his research has focused on the patenting of biotechnology and the transfer of biotechnology to developing countries.